Peroxides, Superoxides, and Ozonides of Alkali and Alkaline Earth Metals

Il'ya Ivanovich Vol'nov

Head, Laboratory of Peroxide Chemistry
N. S. Kurnakov Institute of General and Inorganic Chemistry
Academy of Sciences of the USSR, Moscow

Translated from Russian by
J. Woroncow
Life Sciences Group
General Dynamics / Convair Division
San Diego, California

Edited by
A. W. Petrocelli
Chief, Chemistry and Chemical Engineering Section
General Dynamics / Electric Boat Division
Groton, Connecticut

PLENUM PRESS · NEW YORK · 1966

Born in 1913, Il'ya Ivanovich Vol'nov is head of the laboratory of peroxide chemistry of the N. S. Kurnakov Institute of General and Inorganic Chemistry of the Academy of Sciences of the USSR in Moscow. He joined the Institute in 1939 and since 1949 he has authored more than 50 articles dealing with the chemistry of the inorganic peroxides, superoxides, and ozonides. Vol'nov served as editor for the proceedings of the 2nd All-Union Conference on the Chemistry of Peroxide Compounds, published by the Academy of Sciences in 1963. He was also editor of T. A. Dobrynina's monograph on *Lithium Peroxide,* published in 1964, and edited a bibliographical index covering the world-wide literature for the period 1956 to 1962 on the chemistry of peroxide compounds (other than hydrogen peroxide) published under the auspices of the library of the Academy of Sciences of the USSR.

ISBN 978-1-4684-8254-6   *ISBN 978-1-4684-8252-2 (eBook)*

DOI 10.1007/978-1-4684-8252-2

Library of Congress Catalog Card Number 66-22125

The original Russian text, first published for the N. S. Kurnakov Institute of General and Inorganic Chemistry of the Academy of Sciences of the USSR by Nauka Press in Moscow in 1964, has been corrected by the author for the English edition.

Илья Иванович Вольнов

**Перекиси, надперекиси и озониды
щелочных и щелочноземельных металлов**

PEREKISI, NADPEREKISI I OZONIDY SHCHELOCHNYKH
I SHCHELOCHNOZEMEL'NYKH METALLOV

PEROXIDES, SUPEROXIDES, AND OZONIDES OF ALKALI
AND ALKALINE EARTH METALS

Editor's Preface

Since the early 1930's, Soviet chemists have played a lead-ing role in the study of unfamiliar oxidation state compounds of the peroxide, superoxide, and ozonide types. Interest in the alkali and alkaline earth metal derivatives is now widespread and diverse, and numerous practical applications of these com-pounds have evolved, ranging from their use as air revitaliza-tion materials in space cabins to their use in compounding semiconductor materials.

Professor Vol'nov is eminently qualified to write this monograph since for many years he has been a leading investi-gator and prolific writer in the field of peroxide, superoxide, and ozonide chemistry. He has succeeded in presenting a lucid and detailed discussion of past work, the present state, and the future potential of this area of unfamiliar oxidation state chemistry.

Of particular interest is Professor Vol'nov's extensive compilation of available thermodynamic, kinetic, and structural data for the alkali and alkaline earth peroxides, superoxides, and ozonides. In addition, he has reviewed the known methods of synthesis, as well as the practical applications for which these compounds are suited.

This monograph will be of interest and value to chemists, not only for the information it imparts, but equally for the information it does not impart, thereby illuminating the re-search paths and investigation which must be undertaken in order to increase our knowledge concerning the chemistry of this important class of chemical compounds.

Chemists studying inorganic free radical mechanisms will find much of significance in this work since nearly all the important free radical and ionic species of oxygen, namely OH, HO_2, HO_3, O_3^-, O_2^-, and $O_2^=$, enter into mechanisms postulated for various reactions of these compounds. In Chapter IV, Professor Vol'nov has presented a concise and excellent compilation of the information available on the hydroperoxyl radical, HO_2.

The reader should keep in mind that the author has reported data as he has found them in the literature, and on the whole has limited critical discussion of reported data to information of which he or his colleagues have first-hand knowledge as a result of direct experimentation.

I express my thanks to Professor Vol'nov for his kind cooperation in the preparation of this English-language edition, and also thank Dr. J. Marriott, Mr. R. Uno, and Dr. A. Capotosto, Jr., for their valuable suggestions and critical review of the edited translation, and to Mrs. R. Hill who typed the manuscript of the English-language translation.

A. W. Petrocelli

Westerly, Rhode Island
July, 1966

Foreword

In recent years, a large number of articles and patents have been published dealing with improved methods for the synthesis and stabilization of hydrogen peroxide and its derivatives. The discovery of new inorganic and organic peroxide compounds has also been reported. The practical application of these compounds is continuously increasing and they are important in widely diversified areas of the national economy.

Much of the experimental material that has been published in various journals and in different languages has been summarized in monographs. Since 1951, several such monographs and reviews dealing mainly with hydrogen peroxide have been published abroad. However, much of the information accumulated in the last decade on production methods; properties and application of metal peroxides, superoxides, and ozonides; peroxides and their salts; peroxide complexes; and peroxyhydrates is not adequately discussed in these works.

A brief monograph by W. Wood,[1] published in 1954, gives only a summary description of the production methods, properties, and applications of hydrogen peroxide.

A monograph by W. Machu,[2] the first edition of which has been translated into Russian, was the most complete review available of the production methods, properties, and uses of hydrogen peroxide and its derivatives until the publication of

---

[1]W.S. Wood, Hydrogen Peroxide, The Royal Institute of Chemistry, lectures, monographs, and reports, London (1954).

[2]W. Machu, Das Wasserstoffperoxyd und die Perverbindungen, 2nd edition, Springer Verlag, Vienna (1951).

a book written by W. Schumb, C. Satterfield, and R. Wentworth.[3] The Machu monograph was a supplement to a previously published monograph by O. Kausch,[4] which was more of a bibliography than a review. In his monograph, W. Machu was concerned mainly with hydrogen peroxide. Out of 400 pages, only approximately 100 pages are devoted to peroxide compounds other than hydrogen peroxide. Moreover, many of the data are obsolete. One of the main shortcomings of the monographs written by Machu and Kausch is that the authors are not specialists on peroxide compounds. For this reason they did not present a critical evaluation of the data, nor did they present a creative approach in explaining the theoretical foundations underlining the developments in our knowledge of the chemistry of peroxide compounds.

A discussion of oxygen, ozone, and hydrogen peroxide by P. Pascal is found in the 13th volume of a handbook of inorganic chemistry published in 1960.[5] All aspects of the chemical and physical methods concerning qualitative and quantitative analysis of these substances are thoroughly reviewed in the 6th volume of a handbook of analytical chemistry published in 1953.[6] A review concerning inorganic peroxides has been recently published by N. Vannerberg in which the structure of these compounds is stressed.[7]

The aforementioned monograph by W. Schumb et al., the Russian translation of which appeared in 1958, covers more than 2500 literature sources published up to mid-1954. However, it is almost completely dedicated to hydrogen peroxide. Other inorganic peroxides, the industrial significance of which is just as important and which are of considerable scientific interest, are treated in a brief space of approximately 30 pages, and the discussion of these compounds is limited to information drawn from only 123 literature sources, nine of which appeared in 1953.

[3]W. Schumb, C.N. Satterfield, and R.L. Wentworth, Hydrogen Peroxide, Reinhold, New York (1955).

[4]O. Kausch, Das Wasserstoffsuperoxyd, Knapp Verlag (1938).

[5]P. Pascal, Nouveau Traité de Chimie Minerale, Vol. XIII, Paris (1960).

[6]Handbuch der analytischen Chemie, Vol. VI, Springer, Berlin (1953).

[7]N.G. Vannerberg, Progress in Inorganic Chemistry, Vol. 4, Interscience, New York (1962).

To fill the gap, the author of this monograph has systematized and summarized information from Soviet and foreign chemical journals and patents published from 1950 through 1962. This monograph is devoted to one of the most important branches in the chemistry of inorganic peroxide compounds, namely, the synthesis and properties of peroxides, superoxides, and ozonides of alkali and alkaline earth metals. The alkali and alkaline earth peroxides and superoxides are already economically important compounds which are widely used in various branches of our national economy, as well as that of foreign countries. The alkali and alkaline earth ozonides have only recently been produced in the USSR and they are undergoing detailed and extensive studies.

Because of the difference in the structure and properties of the many types of inorganic peroxide compounds, it is difficult to establish specific regularities describing the formation of peroxide compounds by all of the elements of the periodic table until the available information concerning the separate groups is systematized. For this reason, I have decided to cover in this monograph only peroxide compounds formed by alkali and alkaline earth metals, i.e., simple peroxide compounds. A presentation of the information that has been uncovered in the last decade on the synthesis and properties of peroxides, superoxides, and ozonides of the alkali and alkaline earth metals will serve to establish for the reader a good picture of our present state of knowledge concerning peroxide compounds. Complex peroxide compounds will be discussed in a separate study which is being prepared by the author. A significant part of this review is dedicated to research completed in the Laboratory of Peroxide Compounds of the N. S. Kurnakov Institute of General and Inorganic Chemistry of the Academy of Sciences of the USSR. Extensive use has also been made of the bibliographical index[8] published by the author.

I. I. Vol'nov

---

[8]Chemistry of Peroxide Compounds (Except Hydrogen Peroxide), Bibliographical Index of Domestic and Foreign Literature, 1956 — 1962, Library of the Academy of Sciences of the USSR (1963).

Contents

The Present State and Future Investigations in the Field of Inorganic Peroxide Compounds

The chemistry of inorganic peroxide compounds had its start early in the nineteenth century, when A. Humboldt discovered barium peroxide, G. Gay-Lussac and L. Thenard synthesized sodium and potassium peroxides, and Thenard synthesized hydrogen peroxide.

Russian scientists first showed interest in these compounds during the second half of the nineteenth century. Since then, the contributions made by Russian scientists to this important and complex branch of inorganic chemistry have been outstanding. In Russia, the founders of inorganic peroxide chemistry were D. I. Mendeleev (1834-1907), professor of the Petrovskaya Academy; E. B. Schoene (1834-1896); and P. G. Melikov (Melikashvili, 1850-1927).

Mendeleev did not actually perform experimental studies on peroxides. However, in his work "Principles of Chemistry" and in separate articles, he proposed many interesting ideas which paved the way for a more thorough understanding of these compounds. His ideas have since been confirmed experimentally by both Russian and foreign scientists.

Schoene pioneered the experimental studies of peroxides in Russia. His investigations on peroxyhydrates retain their importance even to this day.

Melikov organized the first school on peroxide studies, and together with his students, L. V. Pisarzhevskii, S. M. Tanater,

I. N. Kazanetskii, A. S. Komarovskii, and others, made significant contributions to this field of inorganic chemistry.

Of special importance is the work performed by Academician A. N. Bakh, founder of the theory of slow oxidation, and
the work of Academician L. V. Pisarzhevskii, the prominent
student of P. G. Melikov, who was the first to apply physico-
chemical methods to the study of peroxides. Mention must also
be made of the work on peroxide compounds carried out by
N. N. Beketov and V. V. Kurilov.

Significant progress in this field of chemistry has been
made since the Great October Socialist Revolution. Studies
conducted in schools founded by L. V. Pisarzhevskii at the
Ukrainian Institute of Physical Chemistry, P. G. Melikov at
the Georgian Chemical Institute, A. N. Bakh and I. A.
Kazarnovskii at the L. Ya. Karpov Physicochemical Institute
in Moscow, N. I. Kobozev and E. N. Shpital'skii at the Moscow
University, A. I. Brodskii at the Institute of Physical Chemistry
of the Ukrainian Academy of Sciences, and investigations based
on N. S. Kurnakov's teachings, which have been conducted by
S. Z. Makarov at the Institute of General and Inorganic
Chemistry of the Academy of Sciences of the USSR, as well as
a series of investigations conducted in separate applied institutes, have contributed to the solution of many problems dealing
with the synthesis, properties, and structure of inorganic
peroxides.

Today, hydrogen peroxide and its derivatives find wide
application in various fields of the national economy. Use is
made of these compounds in bleaching and dyeing processes
of natural and synthetic fibers; in bleaching of wood cellulose,
soap, fats, furs; as components in detergent powders and
synthetic wash products; in inorganic and organic synthesis;
in the food industry — in canning of different products and in
bread baking; in foam plastics production; as initiators of
polymerization processes; in construction products — production of porous concrete; in agriculture — as ripening acceler-
ators of seeds and as insecticides; in medicine and in the
cosmetic industry; in air revitalization in closed spaces and in

isolated-type respiration apparatus; in portable oxygen re-
generators; in pyrotechnology; in extracting certain metals
from ore concentrates; in production of semiconducting materials used in thermogenerators; in treatment and etching of
metallic surfaces; in vulcanization processes of butyl rubber;
as additives to diesel fuel; and in liquid rocket engines.

The historical development of inorganic peroxide chemistry
can be divided into four periods. The first period extends from
1818 (the discovery of hydrogen peroxide by L. Thenard) to
1869 (the formulation of the Periodic Table by D. I. Mendeleev).
This period is characterized by the wide-ranging investigations conducted by Thenard and his co-workers concerning the
reaction of "oxidized water" which resulted in the development
of a whole series of peroxide derivatives. During this same
period, studies were carried out on the reaction of gaseous
oxygen with metals which resulted in the discovery by A.
Garkur of potassium superoxide, first of a new class of
peroxide compounds, which are not derivatives of hydrogen
peroxide. In this same period, a more precise determination
of the sodium peroxide structure was made.

At the time Mendeleev formulated the Periodic Table,
only 19 elements were known to form peroxide compounds.
Mendeleev was especially interested in inorganic peroxides,
because they did not adhere to the physical concepts predomi-
nant in that era. Their properties did not follow the then generally accepted rule according to which elements, if placed
in the order of their increasing atomic weight, would form a
series of higher oxides within a given periodic group. It did
not become clear until the discovery of the molecular oxygen
ions O_2^{2-}, O_2^-, and O_3^- in the 1940's, that the "inorganic peroxide compounds" do not violate the general rules of valence,
periodicity, and complex formation. For example, the KO_2
compound, or K_2O_4 as it was known at that time, is not an
exception to the general rule (according to which the valence
of potassium must be unity since it falls in Group I of the
Periodic System) because the compound is actually characterized by the presence of a single-charge molecular ion O_2^-.

Similar examples are Na_2O_2, which is characterized by the presence of the molecular ion O_2^{2-}, and KO_3, which contains the molecular ion O_3^-.

The second period in the development of inorganic peroxide compounds can be considered to extend from the discovery of the Periodic Table to the application of physical chemistry to the investigation of peroxide compounds at the beginning of this century. This period includes the classical studies of Melikov and Pisarzhevskii, who uncovered a series of regularities in the formation of peroxyacids; the studies of R. de Forcrand in France dealing with the thermochemistry of inorganic peroxides; and the discovery of sodium perborates and carbonate peroxyhydrates of alkaline metals by Tanater. Also, during this period, a new type of peroxide compound was obtained for the first time (peroxyacids and their salts) by the use of electrochemical methods. M. Berthelot synthesized peroxydisulfuric acid, and E. Konstam and A. Henson produced potassium peroxydicarbonate. During this period, methods for industrial production of sodium peroxide and hydrogen peroxide were developed.

The third period is characterized by the extensive studies, both in the USSR and abroad, of the structure, properties, and bond characteristics of peroxide compounds. This period includes the work of Kazarnovskii and his co-workers concerning the structure of a series of peroxide compounds, his discovery of sodium superoxide, and the fundamental investigations carried out by the Canadian scientist Otto Maas and his co-workers concerning concentrated hydrogen peroxide. During this period, which covers the time between the two World Wars, inorganic peroxide compounds began to find wide application in the national economy.

The application during World War II of hydrogen peroxide in German V-2 rockets and in submarine power plants, the necessity to supply oxygen for crews confined in closed spaces and in mines (which can be effectively accomplished by utilizing certain peroxide compounds which liberate oxygen upon reaction with carbon dioxide), the use of certain peroxide compounds in the production of semiconducting materials or as initiators of polymerization processes and in organic synthesis, and the formation of peroxides in radiolysis processes taking place in homogeneous nuclear reactions all necessitated continuing thorough investigation of inorganic peroxide compounds.

Such investigations have been directed mainly to the search for inexpensive methods for the synthesis of hydrogen peroxide in pure and concentrated forms and for the synthesis of other peroxide compounds with high oxygen content.

Significant progress has been made in the Soviet Union in the post-war period in both the theoretical and applied fields. Kazarnovskii made important contributions to this progress as a result of his studies concerning the synthesis, properties, and reactivity of sodium superoxide and of the important new class of peroxide compounds—the inorganic ozonides. It is also appropriate to mention in this context the work of Brodskii, who was the first to establish the mechanism for the formation and disintegration of peroxide compounds via studies involving the use of the heavy oxygen isotopes. Also important was the work of A. M. Gurevich concerning the nature of uranium peroxide compounds, and the work of Kobozev on the synthesis of hydrogen peroxide by electrical discharge. Shpital'skii and his co-workers have made important studies of the catalytic decomposition of hydrogen peroxide. And at the Laboratory of Peroxide Compounds of the Institute of General and Inorganic Chemistry of the Academy of Sciences of the USSR, Makarov has directed many important physico-chemical studies of peroxide compounds.

Abroad, the chemistry of inorganic peroxide compounds is being developed mainly in universities and in private scientific research institutes in the United States, Great Britain, and West Germany. In the United States, for example, systematic studies of the properties of inorganic superoxides, especially the thermodynamic properties of these compounds, have been carried out by Professor Kleinberg at the University of Kansas and Professor Margrave at Wisconsin University.

Extensive investigations were conducted at the Massachusetts Institute of Technology, under the direction of Professor C. Satterfield, on the properties of concentrated hydrogen peroxide solutions. Professor Satterfield is the author of the aforementioned monograph. At the Illinois Institute of Technology, Professor I. G. Solomon directed work on the synthesis of inorganic ozonides. Research on peroxides is also being carried out in the scientific research laboratories of DuPont and FMC Corporation in the United States, and Laporte in Britain. These laboratories have modern equipment and large teams of specialists. Interesting work is being conducted in Canada at Laval University by Professor P. Giguère. His studies have been primarily concerned with concentrated hydrogen peroxide, deuterium peroxides, and the HO_2 radical. Existence of this radical has been recently established by American scientists. In recent years, systematic x-ray studies of inorganic peroxides have been carried out by G. Föppl in West Germany and by N. Vannerberg and R. Shtomberg in Sweden. Original investigations on the synthesis of alkali metal superoxides during self-oxidation of alcoholates are being carried out in France by Professor A. Le Berre. Also in France, Professor C. Izore is studying the use of ion-exchange resins in the synthesis of hydrogen peroxide. In Hungary, spectroscopy is being applied to the study of inorganic peroxides by Professor A. Zimon; and Professors L. Chani, E. Pungor, and F. Sholimoslu are investigating analytical problems of peroxide compounds. Important work is also in progress in East Germany.

A series of conferences on peroxide compounds took place in the years from 1953 through 1961. A symposium on inorganic peroxides, superoxides, and peroxyhydrates was held in Philadelphia in 1953,* and a symposium on peroxide reaction mechanisms was held at Brown University in 1960.† In 1961,

---

*Papers before Symposium on Inorganic Peroxides, Superoxides and Peroxyhydrates, Philadelphia, Pennsylvania (1953).
†Peroxide Reaction Mechanisms, edited by J. O. Edward, New York, Interscience Publishers, Inc. (1962), 245 pp.

the British Society of Industrial Chemistry sponsored an international symposium on peroxide compounds.* A general meeting dedicated to the chemistry of inorganic and organic peroxide compounds was held in Moscow at the end of 1961.† The papers presented at the meeting showed that Soviet scientists have made significant contributions to this field of chemistry. However, the meeting also pointed out the areas of study which require a greater effort. Much more work must be done in the area of thermodynamics and kinetics. Also, a significant lag was noted in the field of analytical chemistry of peroxides and in x-ray structural investigations. To date, no really thorough investigations have been conducted, either by Soviet or foreign scientists, on the reactivity of inorganic peroxide compounds. It was established at that meeting that inorganic chemists working on peroxides must concentrate their efforts on the synthesis of new superoxides and ozonides especially rich in oxygen. Such research should include the application of superhigh pressures of oxygen and ozone, concentrated solutions of hydrogen peroxide in nonaqueous solvents, and electrochemical methods. Physical chemists have an important role to play by determining the thermodynamic, kinetic, mechanistic, and structural properties of peroxide compounds.

In the applied field, it is necessary to concentrate on the sound development of continuous, efficient, and economical production methods of inorganic peroxide compounds and to increase the utilization of peroxides in the national economy.

The ever-broadening scientific investigations which are being conducted in the USSR in the field of peroxide compounds will help to solve one of the problems facing the national economy — the development of "little chemistry."

---

*Chem. and Ind., 1962(1):4; 1962(2):54 (1962).
†Chemistry of Peroxide Compounds. Moscow, Izd. Akad. Nauk SSSR (1963), 315 pp.

Chapter One

Classification and Nomenclature of Inorganic Peroxide Compounds

Inorganic peroxide compounds are commonly described as compounds whose structures include the peroxo-group, $-O-O-$.

For a long time, the presence of this group—the so-called oxygen bridge—was considered satisfactory for defining an "inorganic peroxide compound." Thorough studies of the structure and properties of this wide class of compounds, which began in the 1930's with the active participation of Soviet scientists, have shown that this definition includes at least nine groups which differ in the type of bond between the oxygen atoms in the bridge or the type of bond between the oxygen bridge itself and the element forming a given peroxide compound. Thus, the "oxygen bridge" concept is no longer considered a satisfactory method for classifying peroxide compounds. Certain authors, including F. Hein [1], classify all inorganic peroxides as complex compounds and divide them into three groups: polyoxides, peroxyacids, and peroxyhydrates. However, the polyoxide group includes compounds whose crystal lattice consists of metallic ions (mainly the Ia and IIa elements of the periodic table) and molecular anions O_2^{2-} with a $[:\ddot{O}:\ddot{O}:]^{2-}$ structure [2, 3], O_2^- with a $[:O \text{---} O:]^-$ structure [3, 4], and O_3^- with a $\left[:\underset{100°}{O:}\diagdown\overset{\ddot{O}}{}\diagup O:\right]$ structure [5].

The electronic structures of the O_2^{2-} and the O_2^- ions are depicted in Fig. 1. The O_2^{2-} ion has an even number of electrons, 14, and the symbol for the ground state of this ion is $^1\Sigma$, which is in agreement with the diamagnetism and absence of color which characterize compounds containing this ion. The

9

Fig. 1. Electron structure of O_2^{2-} and O_2^- ions.

molecular orbital notation [6] for the O_2^{2-} ion is $[KK(z\sigma)^2$ $(y\sigma)^2(x\sigma)^2(w\pi)^4(v\pi)^4]$. The O_2^{2-} ion represents an ellipsoid of rotation with the major axis 4.19 A long. The radius of the O^- ion in the O_2^{2-} ion is 1.35 A. The O–O bond length is 1.49 A. The major semiaxis of the O_2^{2-} ion is, consequently, equal to 2.09 A. The minor semiaxis (or transverse radius) is 1.23 A [7, 8]. The bond dissociation energy of O–O in H_2O_2 is 1.5 eV, which is approximately 35 kcal. From the Born–Haber cycle calculations, the heat of the reaction was determined to be: $O_2 + 2e \rightarrow O_2^{2-}$, 175 ± 15 kcal [10]. The determination of this quantity by other methods has resulted in a value of 110 kcal which is considered a more realistic value than the Born–Haber value since the peroxide lattice energy values used in the Born–Haber calculations were only approximate.

The O_2^- ion has an odd number of electrons, 13, and the basic term is πg. According to the molecular orbital notation [6], the O_2^- ion structure is expressed by the formula $[KK(z\sigma)^2$ $(y\sigma)^2(x\sigma)^2(w\pi)^4(v\pi)^3]$, denoting three extra bonding electrons, which account for the paramagnetic and color characteristics of compounds containing this ion. The O_2^- ion may also be represented as an ellipsoid of rotation with a more elongated large axis, the length of which is 4.55 A. The generally accepted value for the radii of the oxygen atoms in the O_2^- ion is 1.28 A [8]. Subsequent investigations [12] indicated that the value of the radii falls in the range 1.32-1.35 Å. The length of the large semiaxis of the O_2^- ion is 2.27 Å. The length of the small semiaxis (or transverse radius) is 1.52 A [4]. The electron affinity of the oxygen molecule is 0.87 ± 0.13 eV, or 20 ± 3 kcal [13]. The energy required for dissociation of the O_2^- ion to $O^- + O$ has been estimated as 2.99 eV, or approximately 69 kcal, based on the interpretation of the Raman spectra of KO_2 [9].

The O_3^- ion has an odd number of electrons, 19, and compounds including this ion are paramagnetic and colored. The O–O bond length in the O_3^- ion is 1.19 A [14], whereas in oxygen and in ozone it is 1.207 and 1.278 A, respectively [5].* The O–O–O bond angle is 100° [10]. The value for the electron affinity of the ozone molecule has been estimated to be 66.5 kcal [15].

Metallic atoms are bonded to the oxygen bridge via ionic bonding, and it is convenient to differentiate between simple and complex inorganic peroxide compounds.

Simple peroxide compounds include the above mentioned compounds and hydroperoxides. The hydroperoxides are characterized by the presence of hydroperoxyl ion HO_2^- and are represented by the general formula EOOH. In these compounds, which include hydrogen peroxide, the peroxo-group is ionically bonded to the element and covalently bonded to the hydrogen atom. The ionic nature of hydroperoxides has been experimentally established for the NH_4OOH compound [19, 20]. It is also confirmed by the fact that hydrogen peroxide is a weak acid with K = $(H^+) (HO_2^-)/(H_2O_2)$ = $2.24 \cdot 10^{-12}$. It should be noted that certain hydroperoxides can be classified as peroxyhydrates, and others as peroxyacids. For example, NaOOH can be represented as a dimer $Na_2O_2 \cdot H_2O_2$, whereas $(OH)_3$ TiOOH can be considered a peroxyacid.

Complex peroxide compounds include peroxides in which the peroxo-group as such (or in the form of H_2O_2 and HO_2) is bonded to the element by a covalent bond rather than an ionic bond. Complex peroxide compounds also include the addition compounds formed with crystallized hydrogen peroxide.

---

*It was established in 1962 [16, 17] that in the compound F_2O_2 fluorine atoms are joined through oxygen atoms, i.e., F–O–O–F. However, this compound does not show peroxide properties. The O–O bond length is 1.22 A, which is not significantly different from the bond length in the oxygen molecule. At -133°C, F_2O_2 (melting temperature -64.3°C) explodes when brought into contact with ice [18].

The characteristic properties of peroxide compounds, both simple and complex, are: the formation of hydrogen peroxide upon reaction with dilute acid solution, the liberation of oxygen as a result of thermal decomposition, and the liberation of oxygen upon reaction with water and other chemical agents. Other inorganic compounds which could serve as oxygen sources, for example, nitrates, chlorates, perchlorates, permanganates, and certain oxides [21], do not react with water to liberate oxygen; however, they can be made to liberate oxygen by the use of suitable catalysts and high temperatures. Potassium dichromate, manganese dioxide, and lead dioxide liberate oxygen only when caused to react with concentrated sulfuric acid solutions. The liberated superoxide oxygen is generally called "active."

Further classification is possible by dividing the simple inorganic peroxide compounds into four groups: 1) hydroperoxides, characterized by the HO_2^- ion; 2) peroxides, characterized by the O_2^{2-} ion; 3) superoxides, characterized by the O_2^- ion; and 4) ozonides, characterized by the O_3^- ion. The difference in the nature of the bond between the oxygen atoms in the $-O-O-$ bridge as it exists in peroxides and hydroperoxides and superoxides and ozonides becomes apparent when one studies the reactions of these compounds with water. The hydrolysis of peroxides and hydroperoxides proceeds as follows:

$$M_2O_2 + 2H_2O \rightarrow 2MOH + H_2O_2$$

and

$$MOOH + H_2O \rightarrow MOH + H_2O_2$$

(where M = metal). The formation of hydrogen peroxide upon hydrolysis of peroxides and hydroperoxides clearly indicates the relationship of these compounds to hydrogen peroxide. For this reason, peroxides and hydroperoxides are considered as being derived from hydrogen peroxide by replacement of one or both the hydrogen atoms by metal atoms.

The hydrolysis of superoxides proceeds with the formation of the unstable intermediate HO_2 radical, hydroperoxyl, i.e.,

$$2MO_2 + 2H_2O \rightarrow 2MOH + 2HO_2$$

$$2HO_2 \rightarrow H_2O_2 + O_2$$

$$2MO_2 + 2H_2O \rightarrow 2MOH + H_2O_2 + O_2$$

As a result, $2/3$ of the active oxygen, which is called "superoxide oxygen," is liberated in the form of O_2 and is determined gasometrically, and $1/3$ of the active oxygen, which is called "peroxide oxygen," is liberated in the form of H_2O_2 and can be determined by permanganate titration in acid solution. The hydrolysis of ozonides proceeds as follows:

$$MO_3 + H_2O \rightarrow MOH + OH + O_2$$

It is to be noted that hydrogen peroxide is not a product of this reaction. In reactions taking place during hydrolysis of superoxides and ozonides, I. A. Kazarnovskii noted that one can consider superoxides as derivatives of the HO_2 radical and ozonides as derivatives of the hypothetical HO_3 radical.

The term peroxide, as used today, was made part of the Russian nomenclature by D. I. Mendeleev. Mendeleev made note of differences in the chemical reactivity of MO_2 peroxides and simple oxides of the same composition. He wrote that "the term peroxide loses its specificity if it is used to describe not only H_2O_2, Na_2O_2, and BaO_2, but also MnO_2 and PbO_2" [22]. Prior to the appearance of Mendeleev's "Principles of Chemistry," it was common practice to describe all such compounds as peroxides. Unfortunately even to this day, some chemists, both in the USSR and abroad, continue to classify as peroxides the oxides of four-valent lead and manganese in whose molecules the oxygen is actually in the form of an O^{2-} ion, as well as N_2O_4, in which case the N–O bond is covalent.

The word hydroperoxide is taken from the German term "hydroperoxyd" [23].

The word superoxide became part of the Russian nomenclature in 1940 [24], and is more suitable than the previously used term "tetraoxide" [25], since it is now firmly established [3, 4] that these compounds do not contain the O_4^{2-} ion. During recent years, the use of the term "superoxide" has become firmly established in the Russian chemical literature [26-29]. In 1947, the term "dioxide" was proposed [30] to correspond to usage in the German literature [31]. Later, the term "superoxid" was proposed [32] in order to correspond to the English "superoxide," first suggested by E. Neuman [33], the French "superoxyde," the German "superoxyd," and the Italian "superossido" [34]. In 1957, the International Union of Pure and Applied Chemistry officially adopted the terms "hyperoxide" and "hyperoxyde" [35, 36] as suggested by H. Remy [37].

The term "inorganic ozonide" was introduced by Kazarnovskii [38], and it has been accepted by the International Union on Pure and Applied Chemistry [35, 36].

Complex peroxide compounds can be divided into five groups:

1. Peroxyacids and their salts. In these compounds the peroxo-group is part of a complex anion. Within this category, one can distinguish between mononuclear and multinuclear peroxyacids and their salts. Generally, in these compounds the coordinating atoms are elements of the IV-VI group — nonmetals C, N, P, S, or metals Ti, V, Cr, Mo, W, etc. Peroxymonosulfuric acid

$$\begin{bmatrix} O{\diagdown}{\atop}{\diagup}O-O \\ {}S{} \\ O{\diagup}{}{\diagdown}O \end{bmatrix} \begin{matrix} H \\ \\ H \end{matrix}$$

is an example of a mononuclear peroxyacid, and peroxydisulfuric acid is an example of a multinuclear peroxyacid

$$\begin{bmatrix} O{\diagdown}{\atop}{\diagup}O-O{\diagdown}{\atop}{\diagdown}O \\ {}S{}S{} \\ O{\diagup}{}{\diagdown}O{}O{\diagup}{}{\diagdown}O \end{bmatrix} \begin{matrix} H \\ \\ H \end{matrix}$$

The O–O bond length in a peroxo-group of peroxyacid derivatives, for example in the $(NH_4)_2S_2O_8$ compound, is 1.46 A [39], which is less than the bond length in the O_2^{2-} ion (1.49 A).

2. Peroxide complexes which are neither peroxyacids nor their derivatives. These may also be divided into mononuclear and multinuclear compounds.

The mononuclear peroxide complexes can further be divided into three subgroups: peroxycomplexes, perhydrocomplexes, and hydroperoxycomplexes. The peroxycomplex subgroup includes compounds in which the peroxo-group is in the internal sphere, for example, $[(UO_2)_2(O_2)_2(H_2O)_8]$. The perhydrocomplexes contain hydrogen peroxide molecules in the internal sphere, for example, $[Fe(H_2O)_5(H_2O_2)]^{3+}$, while the hydroperoxycomplexes are characterized by the presence of the hydroperoxy radical HO_2, for example, $K_4[(UO_2)_2(O_2)_3 (HO_2)(OH)H_2O]$.

Among the multinuclear peroxide complexes we may list:

$$O_2Zr \overset{\displaystyle O—O}{\underset{\displaystyle SO_4}{<\quad>}} ZrO_2; \quad H_4[Cl_5Re–O–O–ReCl_5] \cdot 2H_2O$$

and

$$[(NH_3)_5Co–O–O–Co(NH_3)_5]NO_3)_5$$

In the last compound, the O–O bond length is 1.45 ± 0.06 A [40].

3. Peroxide hydrates. For example, $Na_2O_2 \cdot 8H_2O$.

4. Peroxyhydrates — molecular compounds. For example, $CaO_2 \cdot 2H_2O_2$ and $K_2CO_3 \cdot 3H_2O_2$. Peroxide peroxyhydrates are closely related to the KF·HF type compounds [41]. Ammonia and certain organic compounds having base properties also form molecular compounds containing H_2O_2.

5. Peroxyhydrate hydrates — molecular compounds containing crystallized water and crystallized hydro-

gen peroxide, for example, $BaO_2 \cdot H_2O_2 \cdot 2H_2O$ and $Na_2SO_4 \cdot 0.5H_2O_2 \cdot H_2O$.

The term "peroxyacid" is an English word and has been recommended by the International Union of Pure and Applied Chemistry [34]. The use of this term in the Russian literature, instead of the term "nadkislota" as suggested by Mendeleev [42], is based on the following reasons.

In Russian nomenclature the compound $H_2S_2O_8$ is designated as supersulfuric acid. However, sulfuric acids in Russian literature are referred to as "sulfuric" and "sulfurous" acids. The terms indicate that in the first compound sulfur is hexavalent, whereas, in the second compound it is quadrivalent. The term "supersulfuric acid" would suggest that in this compound the valence of sulfur exceeds six, which is not realistic. The compound H_2SO_5 is designated as monosupersulfuric acid. Based on this nomenclature it has been pointed out [43] that it would be logical to think that the prefix "mono" indicates the number of peroxo-groups in the compound. Actually, in this case, mono indicates the number of sulfur atoms. The compound $H_2S_2O_8$ contains one peroxo-group, and two atoms of sulfur. However, the term supersulfuric does not indicate this. The use of the words "peroxymonosulfuric" and "peroxydisulfuric" for H_2SO_5 and $H_2S_2O_8$ compounds eliminates any misunderstanding. If in any other peroxyacid the number of peroxo-groups is more than one, as, for example, in the peroxychromic acid H_3CrO_8, then it should be designated as tetraperoxymonochromic acid.

With respect to the nomenclature of peroxyacid salts, it should be noted that the continued use in the chemical literature of the designations persulfates, percarbonates, perborates, etc. can lead to confusion in view of the fact that in the case of certain oxygen-containing acids, such as perchlorates, permanganates, perrhenates, etc., the prefix "per" indicates higher degree of oxidation of the element rather than the presence of the peroxo-group. Certain salts of oxygen-containing acids, for example, perchlorates and permanganates, liberate oxygen upon thermal decomposition. The decomposition, however, unlike the thermal decomposition of peroxyacid derivatives, proceeds with the change of valence of the cation within the complex. For example, in the dissociation process of potassium peroxydisulfate $K_2S_2O_8 \rightarrow K_2SO_4 + SO_3 + 0.5O_2$, the sulfur valence remains six, whereas, in the dissociation of potassium perchlorate or permanganate $KClO_4 \rightarrow KCl + 2O_2$ and $2KMnO_4 \rightarrow K_2MnO_4 + MnO_2 + O_2$, the chlorine valence changes from +7 to -1, and that of manganese from +7 to +6 and +4. For peroxyacid salts, the proper terms are peroxysulfates, peroxycarbonates, peroxyborates, etc. The general term "peroxide complexes" is appropriate.

TABLE I
Classification and Nomenclature of Inorganic Peroxide Compounds

Type	Group			Formula
Simple	hydroperoxides			MOOH
	peroxides			M_2O_2
	superoxides			MO_2
	ozonides			MO_3
Complex	peroxyacids and their salts	mononuclear		$[EO_4]^{2-}$; $[EO_5]^{3-}$; $[EO_5]^{2-}$
		multinuclear		$[E_2O_6]^{2-}$; $[E_2O_8]^{4-}$; $[E_2O_8]^{2-}$
	peroxide complexes	mono-nuclear	peroxy	$[M(O_2)_x(A)_y]^{n+}$
			perhydro ...	$[M(H_2O_2)_x(A)_y]^{n+}$
			hydroperoxy .	$[M(HO_2)_x(A)_y]^{n+}$
		multinuclear		$[(A)_yMOOM(A)_y]^{n+}$
	peroxide hydrates			$M_2^iO_2 \cdot xH_2O$
	peroxyhydrates	of peroxides		$M_2^iO_2 \cdot xH_2O_2$
		of salts		$M_n^iEO_m \cdot xH_2O_2$
	peroxyhydrate hydrates	of peroxides		$M_2^iO_2 \cdot xH_2O_2 \cdot yH_2O$
		of salts		$M_n^iEO_m \cdot xH_2O_2 \cdot yH_2O$
Metallo-organic	hydroperoxides			$R_nM(OOH)_n$
	substituted hydroperoxides			$R_nM(OOR)_n$
				$M(OOR)_n$
	peroxides	homoradical		R_nMOOMR_n
		heteroradical		R_nMOOMR_n
		heteroelementary		R_nMOOCR_n

NOTE: M = metal, E = element, A = addend, R = organic radical.

The term "peroxyhydrate" was also recommended by the International Union of Pure and Applied Chemistry [34]. It is preferred over the term "perhydrate," which appeared in the Soviet literature [23, 44] as a translation of German "per-hydrat," or the term "hydroperoxate" suggested by American scientists [45].

Of particular interest is a new class of peroxides — the metallo-peroxyorganic compounds. Their composition consists of one or more metal atoms, peroxo-groups, and organic radicals [46].

The methods available for the synthesis of the various inorganic peroxide compounds vary widely. Peroxides are commonly synthesized via the reaction of oxides with oxygen, or hydroxides with hydrogen peroxide solutions. Superoxides are usually synthesized via the combustion of metals in an oxygen atmosphere at atmospheric or higher pressures. Ozonides are produced in a reaction of ozonized oxygen with hydroxides or with superoxides. Peroxyacids and peroxyacid salts of typical elements are generally synthesized by electrolytic methods. Peroxyhydrates are obtained exclusively from crystallization in hydrogen peroxide solutions. Metallo-peroxyorganic compounds are synthesized via a reaction of metallo-organic compounds with peroxides or by oxidation with oxygen.

Table I shows the classification and nomenclature of inorganic peroxide compounds as suggested [47–49] by the author and supplemented with the latest data.

REFERENCES

1. F. Hein. Chemische Koordination Lehre, Leipzig, Hirzel Verlag (1950), p. 131.
2. J. D. Bernal, E. Djatlova, J. Kasarnowsky, S. Raikhshtein, and A. Ward. Z. Krist. 92:344 (1935).
3. L. Pauling. The Nature of the Chemical Bond [Russian translation], Moscow, Goskhimizdat (1947), p. 330 [originally published in English by Cornell University Press, New York].
4. V. V. Kasatochkin and V. N. Kotov. Zh. Fiz.-Khim. 8:320 (1936); Dokl. Akad. Nauk SSSR 47:199 (1945).
5. L. Pauling. The Nature of the Chemical Bond, Cornell University Press, New York (1960), p. 354.

6. N.G. Vannerberg. The Formation and Structure of Peroxycompounds of Group IIa and IIb Elements, Goteborg, Uppsala, Almqvist (1959).
7. H. Föppl. Z. Anorg. Allgem. Chem. 291:49 (1957).
8. A.P. Altschuller, J. Chem. Phys. 28:1254 (1958).
9. J.A. Creighton and E.R. Lippincott. J. Chem. Phys. 40:1779 (1964).
10. V. I. Vedeneev et al. Zh. Fiz. Khim. 26:1808 (1952).
11. N.G. Vannerberg. In: Progress in Inorganic Chemistry, Vol. 4, edited by F.A. Cotton, New York, Interscience Publishers, Inc. (1962), p. 137.
12. F. Halverson, Phys. Chem. Solids 23:207 (1962).
13. V.I. Vedeneev et al. Disintegration Energy of the Chemical Bond; Ionization Potentials and Their Relation with Electrons, Moscow, Izd. Akad. Nauk SSSR (1962), p. 207.
14. L.V. Azaroff and J. Corvin. Proc. Natl. Acad. Sci. U.S. 49:1 (1963).
15. G.P. Nikol'skii and I.A. Kazarnovskii. Dokl. Akad. Nauk SSSR 72:713 (1950).
16. A. Jackson. J. Chem. Soc. (1962), p. 4585.
17. J.W. Linnet. J. Chem. Soc. (1963), p. 4663.
18. A.G. Streng. Chem. Rev. 63:607 (1963).
19. A. Simon and K. Keishman. Naturwiss. 42:14 (1955).
20. A. Knop and P. Giguère. Can. J. Chem. 37:1794 (1959).
21. Nouveau Traité de Chimie Minerale, Paris, Masson (1960), p. 174.
22. D.I. Mendeleev. Zh. Russ. Fiz.-Khim. Obshchest. 3:284 (1871).
23. Hydrogen Peroxide and Peroxide Compounds, edited by M. E. Pozin, Moscow, Goskhimizdat (1951).
24. A.F. Wells. The Structure of Inorganic Substances, Moscow, IL (1948), p. 346.
25. I.A. Kazarnovskii. Zh. Fiz.-Khim. 14:326 (1940).
26. K.V. Astakhov and A.G. Getsov. Dokl. Akad. Nauk SSSR 81:43 (1951).
27. T.V. Rode. Dokl. Akad. Nauk SSSR 90:1075 (1953).
28. I.I. Vol'nov. Zh. Neorgan. Khim. 1:1938 (1956).
29. Zh. Russ. Khim., Subject Index for 1953-1954, Vol. 2, Moscow (1959), p. 849.
30. I.A. Kazarnovskii. Zh. Fiz.-Khim. 21:248 (1947).
31. A. Helms and W. Klemm. Z. Znorg. Chem. 241:106 (1939).
32. I.A. Kazarnovskii and L.I. Kazarnovskaya. Zh. Fiz.-Khim. 3:293 (1951).
33. E.V. Neuman. J. Chem. Phys. 2:31 (1934).
34. Extrait des comptes rendus de la 17-me Conference, IUPAC, Stockholm (1953), p. 101.
35. Report of the Commission on the Nomenclature of Inorganic Chemistry, IUPAC, 1957, J. Am. Chem. Soc. 82:5530 (1960).
36. Rapport de la Commission de Chimie Minerale, IUPAC, 1957, Bull. Soc. Chim. France, No. 4:580 (1960).
37. H. Remy. Angew. Chem. 68:613 (1956).
38. I.A. Kazarnovskii. Dokl. Akad. Nauk SSSR 64:69 (1949).
39. W.H. Zachariasen and R.L. Mooney. Z. Krist. 88:63 (1933).
40. N.G. Vannerberg. Acta Cryst. 16:247 (1963).
41. N.G. Vannerberg. In: Progress in Inorganic Chemistry, Vol. 4, edited by F.A. Cotton, New York, Interscience Publishers Inc. (1962), p. 166.
42. D.I. Mendeleev. Zh. Russ. Fiz.-Khim. Obshchest., No. 6, sect. 1:561 (1881).
43. S.M. Gussein-Zade. Azerb. Khim. Zh. 2:126 (1959).
44. S.Z. Makarov and V.N. Chamova. Izv. Akad. Nauk SSSR, Otd. Khim. Nauk (1951), p. 126.
45. W. Schumb et al. Hydrogen Peroxide, Moscow, IL (1958), p. 26.
46. T.G. Brilkina and V.A. Shushunov. Works on Chemistry and Chemical Technology of the Chemical Scientific Research Institute of the Gorkii State University 3:505 (1960).

47. I.I. Vol'nov. Reports on Scientific Research of VKhO im. Mendeleev Members
    No. 1:52 (1955).
48. I.I. Vol'nov. Chemistry of Peroxide Compounds. Moscow, Izd. Akad. Nauk SSSR
    (1963), p. 15.
49. I.I. Vol'nov. Brief Chemical Encyclopedia, Vol. 3, Moscow, Sovetskaya Entsiklo-
    pedia (1964), p. 931.

Chapter Two

Peroxides of the Group One Metals
of the Periodic Table

All alkali metals form peroxides. The peroxides of all the alkali metals, with the exception of lithium, can be synthesized by direct oxidation of the metal with oxygen at atmospheric pressure. The reactivities of sodium, potassium, rubidium, and cesium result from the fact that these metals have large atomic radii and low ionization potentials. Lithium does not share such properties and, consequently, lithium peroxide can be synthesized only by the reaction of lithium hydroxide with hydrogen peroxide solutions. Methods for the synthesis of alkali metal peroxides have been known for a long time. In recent years, research has been directed toward the improvement of these methods and toward more precise determinations of the properties and structure of peroxides and their hydrates and peroxyhydrates.

Sodium peroxide, Na_2O_2, is the only compound in this group that is widely used in the national economy. It is produced on a large scale. Potassium peroxide, K_2O_2, rubidium peroxide, Rb_2O_2, and cesium peroxide, Cs_2O_2, are not very stable and have been produced only under laboratory conditions. At present, lithium peroxide, Li_2O_2, has only limited application despite the fact that the active oxygen content (34.8 wt. %) of this compound exceeds that of all the other metal peroxides and it has high thermal stability.

All the aforementioned peroxides are colorless and diamagnetic when in pure form. Technical sodium peroxide has a slightly yellow color as a result of the presence of traces of sodium superoxide, NaO_2 [1, 2]. All alkali metal peroxides

21

react with moisture and carbon dioxide. For this reason, they must be stored in hermetically sealed containers which should be placed in a cool place at a safe distance from flammable materials. Although alkali metal peroxides are not self-igniting, they do pose a fire hazard when brought in contact with organic materials, such as wood, oil, and paper. They also present a moderate explosion danger when in contact with reducing agents in the presence of moisture [2a], and readily dissolve in water with the liberation of heat. In the latter process they form alkali and hydrogen peroxide, and liberate oxygen gas [3]. They react with dilute acid solutions to form the alkali metal salt and hydrogen peroxide.

The other metals of the first group, copper, silver, and gold, do not form stable peroxides, M_2O_2 (M = metal), either by combustion in oxygen or by reaction with hydrogen peroxide solutions. This is because Cu, Ag, and Au exist in several oxidation states and attempts to synthesize these peroxides have therefore resulted in the oxidation of the metals. Unstable peroxides are formed in the course of the reactions of these ions with hydrogen peroxide; however, they are rapidly reduced to oxides.

Higher oxygen compounds of copper and silver, Cu_2O_3 and Ag_2O_3, are oxides in which the Cu and Ag ions are in the +3 state and are not peroxides [4-6]. Copper and silver do have the tendency to form hydroperoxide type compounds, CuOOH [7] and AgOOH [8, 9], but these are quite unstable. The compounds $CuO_2 \cdot H_2O$ and $CuO_2 \cdot H_2O \cdot H_2O_2$ are found at low temperatures [10] in the $Cu(OH)_2–H_2O_2–H_2O$ system and probably exist only in the presence of hydrogen peroxide solution. $CuO_2 \cdot H_2O$ and $CuO_2 \cdot 0.5H_2O$ compounds have been synthesized via the reaction of $Cu(OH)_2$ with hydrogen peroxide at 0°C by V. Frei. The synthesis of these compounds is thought to occur as a result of the replacement of OH^- ions by O_2^{2-} ions [11]. Products with the composition $CuO \cdot xH_2O_2$ were obtained by the same author [12] via the reaction of CuO with hydrogen peroxide and are considered to be H_2O_2 solution products of copper oxide. This conclusion is more probable and no doubt also applies to the reaction products of $Cu(OH)_2$ with H_2O_2. The data reported in references [11, 13] concerning thermal stability and x-ray phase analysis of these compounds, as well as evidence given to support the existence of anhydrous divalent copper peroxide, CuO_2, are not convincing. As a result of magnetic susceptibility measurements, analysis of reaction products with diluted acids, x-ray structural and diffractometric investigations [14-18], and studies with the oxygen isotope O^{18} [19], it has been demonstrated that silver monoxide, AgO, previously considered as a Ag_2O_2 peroxide similar to Na_2O_2, is an oxide of divalent silver rather than a peroxide. X-ray structural investigations [20, 21] indicate that the AgO lattice also includes Ag^{3+} ions in addition to Ag^{2+}. Peroxide compounds of gold have not been established.

Lithium Peroxide — Li_2O_2

Production Methods. Lithium peroxide was first obtained at the beginning of this century by R. de Forcrand [22]. The process used by de Forcrand involved the reaction of hydrogen peroxide with an alcoholic solution of lithium hydroxide to form the compound $Li_2O_2 \cdot H_2O_2 \cdot 3H_2O$, which was then subjected to vacuum desiccation over phosphorus pentoxide for 35 days, resulting in the formation of lithium peroxide. This method has not been significantly changed, and today lithium peroxide containing 99.6 wt.% Li_2O_2 is obtained [23] in a similar reaction by boiling LiOH (2 g/liter) solution with 30% H_2O_2 solution, followed by washing the $Li_2O_2 \cdot H_2O_2 \cdot 3H_2O$ precipitate in ethyl alcohol and drying over P_2O_5.

In an effort to determine the optimum conditions for the synthesis of lithium peroxide, solubility studies of the LiOH–H_2O_2–H_2O system were carried out at −21, −10, 0, +10, and +30.5°C, and the following solid phases were shown to form in the system: LiOH $\cdot$ H_2O; $Li_2O_2 \cdot H_2O_2 \cdot 3H_2O$; $Li_2O_2 \cdot H_2O_2 \cdot 2H_2O$; $Li_2O_2 \cdot 2H_2O_2$; and $Li_2O_2 \cdot H_2O$ [24-27]. The amounts of hydrogen peroxide in the liquid phase for each of the solid phases in the system are given in Table II.

The amount of Li_2O in the liquid phases of this system

TABLE II

LiOH–H₂O₂–H₂O System*

Solid phase	H_2O_2 content in liquid phase, wt, %				
	+30.5°C	+10°C	0°C	−10°C	−21°C
$LiOH \cdot H_2O$	0–1.63	0–4.78	0–4.59	0–4.57	—
$Li_2O_2 \cdot H_2O$	1.63–6.17	—	—	—	—
$Li_2O_2 \cdot H_2O_2 \cdot 3H_2O$	6.17–27.04	4.78–41.23	4.59–39.50	4.57–40.29	22.96–38.10
$Li_2O_2 \cdot H_2O_2 \cdot 2H_2O$	27.04–58.36	41.23–58.57	39.50–55.50	40.29–54.45	38.00–57.08
$Li_2O_2 \cdot H_2O_2$	58.36–64.57	58.57–64.84	55.50–68.73	54.45–75.66	57.08–83.96

*For more details about this system, see reference [27a].

varies from 7 to 2 wt.% depending on the H_2O_2 concentration, temperature, and the type of solid phase. The existence of the monoperoxyhydrate trihydrate has not been firmly established [23,28].

Studies of the thermal decomposition of $Li_2O_2 \cdot H_2O$ and $Li_2O_2 \cdot H_2O_2 \cdot 2H_2O$ indicate that the first compound is dehydrated at 100-140°C [25] (see Fig. 4a). In the case of the second compund, oxygen of the crystallized hydrogen peroxide and crystallized water is lost in the temperature interval from 85 to 150°C [25, 27].

These investigations serve as a basis for the industrial synthesis of Li_2O_2 from $Li_2O \cdot H_2O$ and result in the production of a material containing up to 95 wt.% Li_2O_2 [25].

In laboratories, lithium peroxide can be obtained via the reaction of concentrated hydrogen peroxide with lithium ethylate or with lithium methylate dissolved in the corresponding alcohol [28, 29]. This method enables one to produce larger quantities of lithium peroxide than are obtainable from aqueous-alcohol LiOH solutions [23] since the solubility of lithium methylate in methyl alcohol is quite high (90 g/liter). According to A. Cohen [28], $LiOOH \cdot H_2O$, rather than $Li_2O_2 \cdot H_2O_2 \cdot 2H_2O$, is the intermediate product in the synthesis process. The material decomposes during the vacuum drying process to yield Li_2O_2. $LiOOH \cdot H_2O$ has a rhombic lattice structure and the parameters are as follows: a = 7.92 A, b = 9.52 A, c = 3.20 A, Z = 4. Density = 1.69 g/cm^3.

A 1960 patent [30] describes a method for the synthesis of lithium peroxide in a methyl alcohol medium. Five hundred grams of anhydrous lithium oxide hydrate is dissolved in 7500 ml of 95% CH_3OH and the solution is filtered. Three hundred milliliters of 27.5% H_2O_2 solution is added to the filtrate with stirring. Upon completion of the reaction (30 min), the lithium peroxide sediment is separated from the mother solution by filtration. It is washed with methyl alcohol and vacuum dried (10 mm Hg) at 90-100°C. The Li_2O_2 yield is 380 g.

The Lithium Corporation of America [31] has developed several methods for the industrial synthesis of lithium peroxide. One of the methods is based on a reaction of nearly stoichiometric amounts of concentrated aqueous hydrogen peroxide and solid lithium oxide hydrate at room temperature. The lithium peroxide formed is separated from the mother solution by filtration at higher pressures or by centrifuging. The precipitate is washed in methyl alcohol and is dried by a stream of CO_2-free air. The end-product contains 95-97% Li_2O_2 and the yield, based on lithium, is 85%.

In another method, a 27-35% hydrogen peroxide solution 1 to 20% in excess of the stoichiometric ratio is added to a saturated, freshly prepared and filtered lithium oxide hydrate solution. The solution is injected into a spray drier with an inlet temperature of 200-300°C. Lithium peroxide is deposited in a collector at 80-150°C.

In a third method, stoichiometric amounts of $LiOH \cdot H_2O$ and 35-50% hydrogen peroxide solution are mixed and poured into boiling toluene or xylene. The solvents form azeotropes with water which are removed by distillation, leaving anhydrous lithium peroxide.

Other methods for the synthesis of lithium peroxide involve the oxidation of lithium amalgam [32] and the reaction of lithium nitrate with sodium superoxide in liquid ammonia at −30°C [33]. However, these methods are of no practical interest because of the low product yield (30-75% Li_2O_2). Attempts to obtain lithium peroxide by oxidizing LiOH with oxygen at high temperatures (500°C) and pressures have not been successful [34].

According to the information in Table II, lithium peroxide can attach crystallized hydrogen peroxide. $Li_2O_2 \cdot 2H_2O_2$ is synthesized by treating $Li_2O_2 \cdot H_2O_2 \cdot 2H_2O$ with 90-95 wt.% hydrogen peroxide solution at −15 to −20°C, followed by drying in vacuum at 0°C [35]. This compound begins to disintegrate at 50-60°C. Thermal studies [25, 27] indicate the presence of an exothermal effect at 100°C, corresponding to the dissociation of the crystallized hydrogen peroxide, and two endothermal effects. The first corresponds to dehydration at 110°C; the second one corresponds to the dissociation of Li_2O_2 to Li_2O and $\frac{1}{2}O_2$ at 310-334°C. Analysis of the product obtained from vacuum drying (10 mm Hg) at 110-120°C of $Li_2O_2 \cdot 2H_2O_2$ indicates the presence of 7-9 wt.% of lithium superoxide mixed with lithium peroxide [36, 37].

Properties of Lithium Peroxide. Lithium peroxide is thermally stable. It does not completely liberate its oxygen unless heated to 315-342°C. Thermal studies of the Li_2O_2–Li_2O system have revealed the existence of an irreversible polymorphous transformation at 225°C, which has been accounted for on the basis that two forms of Li_2O_2 exist: a-Li_2O_2, stable up to 225°C, and β-Li_2O_2, stable up to 280-315°C [38]. The presence of this irreversible polymorphous transformation in Li_2O_2 has been debated [39]. The density of Li_2O_2 is 2.363 g/cm^3 [40]. The unit cell is hexagonal with constants a = 3.142 ± 0.005 A and c = 7.650 ± 0.005 A, c/a = 2.44, and Z = 2. A density of 2.33 g/cm^3 has been determined from X-ray studies. The molecular volume is 19.7 cm^3, and the space group is C_{3h} [41]. Infrared studies of lithium peroxide have been carried out; however, since the O_2^{2-} ion is homopolar, it is inactive in the infrared region [42].

For the reaction $Li + \frac{1}{2}O_2 \rightarrow \frac{1}{2}Li_2O_2$, $\Delta F°_{298} = -68$ kcal, $\Delta H°_{298} = -76 \pm 2$ kcal, $\Delta S°_{298} = (-26)$ eu [43]. In reference [44], values for ΔH and ΔF at 400, 452, and 500°K are presented.

Lithium peroxide reacts with water vapor in the temperature interval 23-300°C with the formation of lithium oxide hydrate [45]. With moist carbon dioxide, lithium peroxide forms the carbonate in the temperature interval 100-200°C [45]. Lithium peroxide does not react with liquid ozone [46].

In recent years, interesting applications have been found for Li_2O_2. It can be sintered at 900°C with oxides of manganese, nickel, cobalt [47-49], zinc, and copper [50] and with solid solutions of transition metal oxides such as $Co_yNi_{1-y}O$ [51] to form binary and ternary compounds which possess semiconducting properties and which find application in thermogenerators. Equations illustrating these reactions are:

$$\tfrac{1}{2}x Li_2O_2 + (1-x)MnO \rightarrow Li_x Mn_{(1-x)}O$$

$$\tfrac{1}{2}x Li_2O_2 + (1-x)Co_y Ni_{(1-y)}O \rightarrow Li_x [Co_y Ni_{(1-y)}]_{(1-x)}O$$

For example, a positive thermoelement is produced by sintering 0.96 mole CuO with 0.07 mole ZnO and 0.01 mole Li_2O_2 at
900°C. The product is then melted and recrystallized [50a] to
a single monocrystal. Lithium peroxide is also used for the
production of anhydrous lithium oxide [29]. A metallo-peroxyorganic compound of lithium, nC_4H_9OOLi, has been synthesized
[52].

Sodium Peroxide – Na_2O_2

Production Methods. Sodium peroxide is obtained by oxidizing
molten metallic sodium in a stream of dry CO_2-free or oxygen-
enriched air at a reaction temperature of about 300–400°C.
The method was introduced by H. Castner [53] at the end of the
last century. Sodium peroxide is still produced by essentially
the same method, although many new types of apparatus have
been introduced [54, 55]. The oxidation of sodium to Na_2O_2
occurs in two steps. First, sodium is oxidized at 150–200°C to
Na_2O, and in the second step Na_2O is oxidized to Na_2O_2 at 350°C
[56, 57]. Figure 2 shows that the Na_2O_2 composition is achieved
in the temperature range of 265–370°C [58].

In practice, the oxidation of metallic sodium to Na_2O_2 can
be effected without isolation of the Na_2O formed in the first step
of the reaction. In recent years, the synthesis of sodium
peroxide has been performed in a sprayer apparatus [54] in
which the dispersed metal is directly oxidized to peroxide.
The sprayer method is also used in the production of potassium
superoxide, KO_2; however, in the production of Na_2O_2 the
degree of dispersion of the metal must be higher because
sodium is less reactive with oxygen [57].

According to a 1953 patent [59], sodium peroxide can also
be produced by oxidizing sodium oxide with a 2:3 oxygen–inert

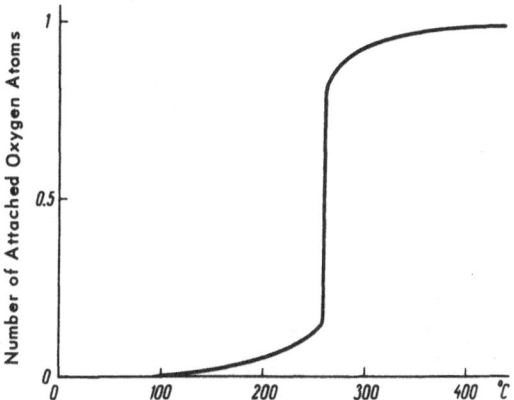

Fig. 2. The relation of reacted oxygen quantity and temperature in the reaction $2Na_2O + O_2 \rightarrow 2Na_2O_2$ [58].

gas (or air) mixture at 350-490°C and 5 atm pressure. The gas phase is periodically renewed until the reaction is completed. Sodium peroxide can also be obtained using the boiling layer method, in which molten sodium is brought in contact with a boiling layer of sodium oxide melted by a stream of CO_2-free air [54]. To obtain a high-quality sodium peroxide product, certain American companies are employing a reduction method by which low-quality sodium peroxide is first reduced to sodium oxide via the following reaction:

$$2Na + Na_2O_2 \rightarrow 2Na_2O$$

The reduction reaction is carried out at 130-200°C, using additions of metallic sodium (in the amount 1-10 wt.% sodium peroxide) in an inert atmosphere containing water vapor. The quantity of water vapor is approximately 0.03-1.3% of the sodium weight. The sodium oxide obtained is then oxidized to peroxide in rotary furnaces at 250-400°C, the peroxide is sifted, and the fraction of the product with particle size −20 ± 40 mesh is removed from the cycle. The +20 fraction is ground, mixed with the −40 fraction, and returned into the cycle. The product obtained in this manner contains 96-98% Na_2O_2 [60-66]. The reduction of sodium peroxide with metallic sodium is also used as an independent method for obtaining pure sodium oxide [58].

A method has been suggested for obtaining sodium peroxide by depositing a layer of molten sodium, approximately 1 mm thick, on the external surface of a steel cylinder. The cylinder is heated to 450-575°C and rotated at a rate so that the deposited layer remains for approximately 15 min in a stream of air. The formed sodium peroxide is then removed from the cylinder [67]. Recently, a new electrolyzer was introduced for dissolving sodium chloride using a molten lead cathode. Sodium vapor formed in the electrolysis process can be oxidized to obtain sodium peroxide [68].

Sodium peroxide is quite corrosive on contact with metals. Therefore, its production is usually carried out in reactors made of nickel alloys coated with graphite [54], and equipped with stirrers made of zirconium [69]. Molten sodium peroxide attacks steel, nickel, platinum, silver, and china crucibles. Decomposition of certain ores with sodium peroxide in a platinum vessel is permissible only at temperatures below 500°C [70]. For this reason, in analytical work requiring the use of sodium peroxide at temperatures greater than 500°C, the use of zirconium crucibles is recommended [71, 72]. It has been established, however, that when a mixture of zirconium and sodium peroxide in a ratio of 1:30 is heated over a bunsen burner flame, the zirconium loses 4% of its weight in 2 hr [73]. Metallic vanadium at the same ratio and temperature is completely dissolved in 2 hr, and titanium is dissolved in 10 min [73]. The weight loss of germanium upon reaction with molten sodium peroxide is 54.2 mg/cm^2 [74].

Figure 3 shows the weight losses, in milligrams, of standard crucibles made of different materials occurring after 10 min of reaction with 5 g of molten sodium peroxide at temperatures over 500°C [75]. Magnesium oxide crucibles have been used to prepare sodium peroxide crystals for x-ray structural studies [76].

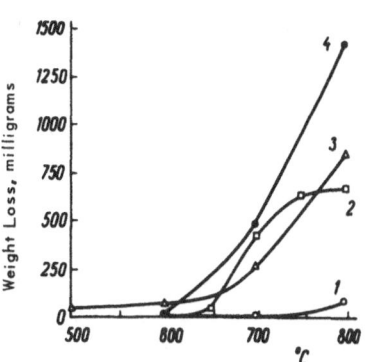

Fig. 3. Relative weight loss of crucibles made of different materials in
reactions with molten sodium peroxide [75]. (1) Zirconium; (2) steel;
(3) china; (4) nickel.

According to a 1953 patent [77], 98.7%-pure sodium peroxide can be obtained by oxidizing a sodium amalgam at 20°C.
The best result is obtained using an amalgam containing 0.1%
Na [32]. However, separation of sodium peroxide from mercury presents tremendous difficulties since sodium peroxide
catalyzes the oxidation of mercury to HgO [78]. For this
reason, specialists [79] believe this method is not suitable for
large-scale production. Oxidation of amalgams containing
0.02–0.05% Na does not produce sodium peroxide [80].

Other reactions for producing sodium peroxide are also
known. Sodium peroxide can be obtained via the reaction
$2Na_2O \rightarrow Na_2O_2 + 2Na$ at 600°C [58]. The oxidation of NaOH
with oxygen at pressures greater than 1 atm at 400°C produces
a product containing only 5 wt.% Na_2O_2 [34]. A low yield of
sodium peroxide is obtained from a reaction involving sodium
nitrate, i.e.,

$$NaNO_3 + 3NaOH + 2Na \rightarrow 3Na_2O_2 + NH_3$$

Quite interesting is the method for obtaining sodium peroxide based on the oxidation of organic derivatives such as
anthrahydroquinone, benzhydrol, fluorenol, and diphenyl-

dihydroanthrol.   An example of the reaction scheme is [81]

$$\text{(structure)} \xrightarrow{+2\,Na} \text{(structure)} \xrightarrow{+O_2} \text{(structure)} + Na_2O_2$$

In the auto-oxidation of organometallic compounds which
do not belong to polynuclear aromatic or olefinic aromatic
hydrocarbons (e.g., metal alkyls), unstable organometallic
peroxides ROOM are formed which, in the presence of an
excess of RM, are reduced to alcoholates. During the auto-
oxidation of metal aryls, peroxides are generally not formed.
The phenols and diaryls are exceptions [82].

Sodium peroxide is also formed during the oxidation of
complexes of organic bases and sodium of the type $(C_6H_5N)_2 \cdot Na$
[83].

Among the methods based on the use of readily oxidizable
organic compounds, the production of sodium peroxide by
oxidizing hydrazobenzene in the presence of sodium alcoholate
has been introduced into industrial practice in a number of
foreign countries [54, 84, 85]. In this method, azobenzene is
reduced, in a medium such as ethyl alcohol, to hydroazobenzene

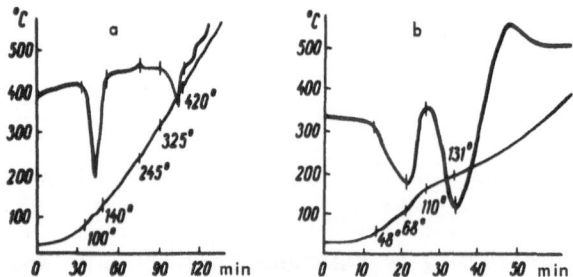

Fig. 4. (a) $Li_2O_2 \cdot H_2O$ thermograph [25]; (b) $Na_2O_2 \cdot 8H_2O$
thermograph [86].

by sodium amalgam with simultaneous formation of sodium ethylate according to the reaction

$$C_6H_5N{=}NC_6H_5 + 2NaHg + 2C_2H_5OH \rightarrow$$

$$C_6H_5NH{-}HNC_6H_5 + 2NaOC_2H_5 + 2Hg$$

Upon oxidation of the hydrazobenzene and sodium ethylate mixture with molecular oxygen, sodium peroxide is formed and separated by filtration. The regenerated azobenzene and alcohol are returned into the cycle:

$$C_6H_5NH{-}HNC_6H_5 + 2NaOC_2H_5 + O_2 \rightarrow$$

$$C_6H_5N{=}NC_6H_5 + 2C_2H_5OH + Na_2O_2$$

Production of sodium peroxide by a method similar to that used for lithium peroxide, i.e., dehydration of the solid phases formed in the $NaOH-H_2O_2-H_2O$ system, $Na_2O_2 \cdot 8H_2O_2$ and $Na_2O_2 \cdot 2H_2O_2 \cdot 4H_2O$, does not seem to be feasible. According to Fig. 4b, sodium peroxide octahydrate melts at ~50°C, and at ~110°C loses its water of crystallization and completely liberates its active oxygen, whereas $Li_2O_2 \cdot H_2O$ can be dehydrated at 140°C without disintegration (Fig. 4a). The compound $Na_2O_2 \cdot 2H_2O_2 \cdot 4H_2O$ liberates active oxygen at 60-65°C from both the bound hydrogen peroxide and from the peroxide [86, 87].

Properties of Sodium Peroxide. Na_2O_2 exists in three modifications: $Q-Na_2O_2$, stable at the temperature of liquid air; $Na_2O_2(I)$, stable up to $512 \pm 1°C$; and $Na_2O_2(II)$, stable from the above temperature to the melting point, 596°C [88, 89].

The density of $Na_2O_2(I)$ is 2.60 g/cm^3 [41]. The unit cell is hexagonal (irregular) with constants a = 6.208 A, c = 4.469 A; c/a = 0.72. According to the work of Margrave et al. [88], the constants are a = 6.207 ± 0.004 A and c = 4.471 ± 0.003 A; Z = 3. Based on x-ray data, the density is 2.62 g/cm^3. The molecular volume is 30.0 cm^3. The space group is D_{3h}^3 [41]. Figure 5

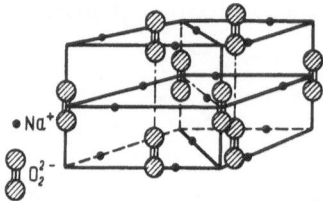

Fig. 5. Na_2O_2 crystal lattice [41].

shows a schematic of the sodium peroxide crystal lattice [41]. The Na_2O_2 lattice energy is estimated to be 532 kcal/mole [90]. Infrared studies of sodium peroxide, as well as lithium peroxide, did not produce specific clues concerning the structure of these substances [42, 91]. The ultraviolet spectrum of technical grade sodium peroxide indicates the presence of sodium superoxide in these products [92].

For the reaction $Na + \frac{1}{2}O_2 \rightarrow \frac{1}{2}Na_2O_2$, $\Delta F°_{298} = -53$ kcal [43], $\Delta H°_{298} = 61.05 \pm 0.5$ kcal [93], $\Delta S°_{298} = -25.4$ eu [43]. Reference [44] includes ΔH and ΔF values for the temperature interval from 298 to 1500°K. The specific heat capacity of sodium peroxide, $C_p°_{298}$, is 21.35 cal/mole–deg [94]. Heat capacity values for higher temperatures are presented in reference [95]. The entropy of Na_2O_2, $S°_{298}$, is 22.6 ± 0.3 eu [92]. Na_2O_2 is diamagnetic, $\chi_m \cdot 10^6 = -24$ [96].

The melting point of sodium peroxide has not been precisely established, although it is known to be somewhat higher than 596°C [95]. Sodium peroxide begins to liberate active oxygen in the temperature range 311–400°C. It rapidly dissociates at 540°C, and at 636°C the dissociation pressure, P_{O_2}, is 1 atm [43]. Thermal studies of sodium peroxide indicate a phase transition at 510°C, which was originally thought to be the melting temperature of Na_2O_2 [97, 98]. However, it is now thought [88, 95] that the thermal effect at 510°C corresponds to the Na_2O_2(I) to Na_2O_2(II) transition.

For the temperature range 15–450°C, the thermal expansion coefficient of sodium peroxide is $(2.84 \pm 0.4) \cdot 10^{-5}$/deg C [88, 89].

The dissociation temperature of sodium peroxide can be significantly reduced by mixing heavy metal oxides with the peroxide.   For example, TiO_2 mixed in a 1:1 ratio with Na_2O_2 reduces the dissociation temperature to 250°C [99], and the oxides UO_3, U_3O_8, and U_3O_7 reduce it to 239, 182, and 300°C, respectively [100].   Between 70 and 330°C, Na_2O_2 and UO_3 form the compound $Na_2O_2 \cdot UO_3$, and in the range of 450 to 500°C, Na_2UO_4 is formed [81].   In a mixture of Na_2O_2 and GeO_2 in a 1:1 ratio, the peroxide dissociates at 269°C and Na_2GeO_4 is formed [100].   Similarly, Sb_2O_3 reduces the dissociation temperature of Na_2O_2 to 300°C with the formation of $NaSbO_2$ [100, 101].   A mixture of Na_2O_2 and CrO_3 in a 1:6 ratio undergoes reaction in the temperature range from 100 to 660°C to form mixtures of the compounds Na_2CrO_4 and $Cr_2(Cr_2O_4)_3$ [99, 100]. MoO_3 reacts with Na_2O_2 in a 1:1 ratio to form $NaO_2 \cdot MoO_3$ in the temperature range from 80 to 320°C, Na_2MoO_4 in the temperature range from 320 to 505°C, and Na_2MoO_7 in the range from 505 to 750°C [101-103]. Bi_2O_3 reacts with Na_2O_2 to form Bi_2O_5 and $Na_2O \cdot Bi_2O_5$ in the temperature range from 305 to 430°C [101], and WO_3 reacts with the peroxide to form $Na_2O_2 \cdot WO_3$ in the 70 to 330°C range, and Na_2WO_4 in the range 330 to 510°C [101, 103].   Between 326 to 600°C, Al_2O_3 reacts with Na_2O_2 to form $Na_n(AlO_2)_n$ (where n = 2 or 3) [101, 104]. Since the thermal studies of Na_2O_2(II), discussed above, were carried out in an Al_2O_3 vessel, some doubt is cast on the interpretation given to the data in reference [89].   The heating of 1:6 mixtures of SiO_2 and Na_2O_2 at 260°C causes the formation of orthosilicates [105].   And in the temperature range from 165 to 400°C, V_2O_5 and Na_2O_2 form $2Na_2O_2 \cdot V_2O_5$ and Na_3VO_4. With Nb_2O_5, the compound Na_2NbO_4 is formed in the 43 to 470°C range; and with Ta_2O_5, the compound Na_3TaO_4 is formed in the range from 310 to 400°C [101].

Sodium peroxide reacts with water to liberate a significant quantity of heat [93], i.e.,

$$Na_2O_2 + 2H_2O \rightarrow 2NaOH + H_2O_2 + 34 \pm 0.3 \text{ kcal/mole}$$

Sodium peroxide can be caused to detonate when mixed, in the presence of moisture, with reducing agents such as $SbCl_3$,

As_2O_3, Sb_2S_5, $Al(CNS)_3$, powdered iron, aluminum, and calcium carbide, as well as finely dispersed sulfur, $AlCl_3$, CH_3COOH, ethyl ether, glycerine, and sugar [106, 106a]. The reaction of Na_2O_2 with DOH is used to prepare sodium deutero-oxide [107].

The determination of the active oxygen content of the alkali metal peroxides is usually accomplished by means of a permanganate titration. Since dissolving the compound in water liberates a significant quantity of heat, and the formed H_2O_2 decomposes in a basic solution to H_2O and $\frac{1}{2}O_2$ [3], the best analysis is obtained by titrating in an acid medium (H_2SO_4) in the presence of $MnSO_4$ [108]. Na_2O_2 contains 20.5 wt.% active oxygen.

Under certain conditions, sodium peroxide can be caused to react with water to form $Na_2O_2 \cdot 8H_2O$, a compound which is produced on an industrial scale. Under laboratory conditions, sodium peroxide octahydrate is obtained via the reaction of sodium hydroxide solutions with hydrogen peroxide at 0°C [86, 109], or by oxidizing hydroxylamine in a 3% alcoholic solution of sodium oxide hydrate [110, 111].

The compound $Na_2O_2 \cdot 8D_2O$ is also known [42]. The $Na_2O_2 \cdot 8H_2O$ existence region in the $NaOH-H_2O_2-H_2O$ system is limited to H_2O_2 concentrations of 5 to 10 wt.%. Sodium peroxide octahydrate is crystallized in the form of hexagonal monoclinic plates. The lattice parameters are: a = 13.52 A, b = 6.46 A, c = 11.51 A. The density is 1.61 g/cm^3. X-ray density is 1.57 g/cm^3 [112].

The industrial method of producing $Na_2O_2 \cdot 8H_2O$ is based on the oxidation of hydrazobenzene and sodium alcoholate in the presence of the necessary amount of water [113, 114] according to the following reaction:

$$C_6H_5NH-HNC_6H_5 + 2NaOC_2H_5 + 8H_2O + O_2 \rightarrow$$

$$C_6H_5N=NC_6H_5 + Na_2O_2 \cdot 8H_2O + 2C_2H_5OH$$

The alcohol and azobenzene are recycled as in the process for obtaining anhydrous sodium peroxide [84]. Using this method, a plant in Kuusankoski, Finland, produced 120 tons per month of sodium peroxide octahydrate in 1959. In 1960, it was reported that the daily production of this plant would be increased to 14 tons [113, 114].

Sodium peroxide octahydrate reacts with a controlled supply of CO_2 at 0-30°C, and with the simultaneous removal of water, $Na_2C_2O_6$ is formed [115].

If stored for extended periods over sulfuric acid in a vacuum desiccator, the crystalline octahydrate gives off six molecules of water and is transformed into the crystalline dihydrate $Na_2O_2 \cdot 2H_2O$ [115a]. The conversion of the octahydrate to the dihydrate proceeds at 20°C and 0.3-1.5 mm Hg. The $Na_2O_2 \cdot 2H_2O$ lattice is tetragonal with parameters a = 8.75 A, c = 3.73 A; space group P4/m [115b].

When sodium peroxide reacts at 0°C with H_2O_2 solutions that are more concentrated than 30 wt.%, it forms sodium peroxide diperoxyhydrate tetrahydrate $Na_2O_2 \cdot 2H_2O_2 \cdot 4H_2O$ [86]. In the reaction of sodium ethylate with concentrated hydrogen peroxide, a compound is formed which is considered by certain authors to be the hydroperoxide NaOOH, whereas other authors believe it to be the peroxyhydrate compound $Na_2O_2 \cdot H_2O_2$. The first classification is supported by the reaction of the compound with carbon dioxide to form the acid salt of peroxycarbonic acid $NaHCO_4$ [116]; the second classification is supported by the differential heating curve which shows an exotherm at 65°C, corresponding to the decomposition of the bound hydrogen peroxide with the formation of $Na_2O_2 \cdot H_2O$ (melting point 285°C) [87].

The $Na_2O_2 \cdot 2H_2O_2$ compound can be produced either by the dehydration of the tetrahydrate [86], or by the reaction of sodium methylate with hydrogen peroxide [116-118]. Sodium peroxide diperoxyhydrate, like sodium peroxide, is produced by the oxidation of hydrazobenzene and a sodium methylate mixture in which the hydrazobenzene : methylate : oxygen ratio is 3:2:3 rather than 1:2:1 [118, 119]. The reaction proceeds as follows:

$$3C_6H_5NH-HNC_6H_5 + 2CH_3ONa + 3O_2 \rightarrow$$

$$3C_6H_5N=NC_6N_5 + 2CH_3OH + Na_2O_2 \cdot 2H_2O_2$$

When $Na_2O_2 \cdot 2H_2O_2$ is heated, the crystallized hydrogen peroxide decomposes at 157°C [117]. Heating $Na_2O_2 \cdot 2H_2O_2$ in vacuum (10 mm Hg, 70-120°C) yields a mixture of sodium peroxide, superoxide, and hydroxide [117, 118]. The unit cell dimensions of $Na_2O_2 \cdot 2H_2O_2$ are: a = 10.74 ± 0.03 A, b = 5.18 ± 0.03 A, c = 8.43 ± 0.05 A; β = 71°12' ± 24'. The density is 2.18 g/cm^3 and the space group is C2/c No. 15 [115b].

In the oxidation of sodium benzhydrolate in benzene or petroleum ether, sodium hydroperoxide and benzophenone are formed in 95% and 98% yields, respectively, according to the following reaction:

$$C_6H_5-CHONa-C_6H_5 + O_2 \rightarrow C_6H_5-CO-C_6H_5 + NaOOH$$

NaOOH is stable up to 70°C, decomposing above this temperature as follows [120]:

$$2NaOOH \rightarrow 2NaOH + O_2$$

Under similar conditions, other aromatic alcoholates also form sodium hydroperoxide, but in lower yields [121].

Sodium peroxide reacts with atomic hydrogen to form sodium hydroxide [122]. In the reaction with moisture and CO_2, sodium peroxide forms sodium hydroxide and sodium carbonate and liberates oxygen. Moisture and CO_2 are absorbed at room temperature and up to 140 ± 10°C. Above this temperature, up to 850°C, the mixture continues to react with CO_2 liberating moisture, while above 850°C, the sodium carbonate initially formed begins to decompose [100, 101]. Sodium oxide hydrate and sodium carbonate are practically always present in sodium peroxide. A study of the processes taking place during the reaction of sodium peroxide with these substances has indicated that the peroxide does not react with completely dry sodium carbonate during heating, and that the reaction with anhydrous sodium hydroxide begins at the melting point of the hydroxide (300-310°C). In the reaction of sodium peroxide with molten hydroxide, oxygen is liberated and up to 15 mole percent of sodium oxide is dissolved. The formed homogeneous phase corresponds to a solid solution of sodium oxide in the hydroxide. The excess quantity of sodium peroxide is decomposed at higher temperatures. In the presence of $NaOH \cdot H_2O$ or $NaOH \cdot 0.5H_2O$ mixtures, or if insufficiently isolated from moisture, sodium peroxide decomposes with the evolution of oxygen at 60 and 170°C, respectively [123]. Sodium peroxide does not react with $NaHCO_3$. At 100°C, the bicarbonate begins to decompose [123].

Sodium peroxide reacts with N_2O_4 at ~ 140°C or at higher temperatures to form the nitrate radical; below 40°C it forms the nitrite radical [124].

The phase diagram of the $NaNO_3-Na_2O_2$ system has been established. It indicates the existence of solid solutions in the sodium peroxide concentration region below 15 wt.% and over 90 wt.% and eutectics at 240°C at a composition of 75% Na_2O_2 [58].

Sodium nitrite reacts with sodium peroxide at 300°C with the formation of nitrate [58]. A mixture of sodium peroxide and ammonium nitrate reacts in liquid ammonia to form hydrogen peroxide and sodium nitrate [125]. According to references [118, 126], sodium peroxide reacts weakly in liquid SO_3 even at 100°C to form Na_2SO_5, $Na_2S_2O_8$, and $Na_2S_2O_7$, according to the following reactions:

$$Na_2O_2 + SO_3 \rightarrow Na_2SO_5$$

$$Na_2O_2 + 2SO_3 \rightarrow Na_2S_2O_8$$

$$Na_2O_2 + 3SO_3 \rightarrow Na_2S_2O_7 + SO_2 + O_2$$

According to reference [127], a sodium peroxide and SO_3 mixture dissolved in SO_2 detonates at 7-10°C, while at -15°C, it reacts with the formation of $Na_2S_4O_{14}$. A sodium peroxide and $Na_2S_2O_3$ mixture detonates in the presence of moisture [128]. With ClO_2, sodium peroxide reacts to form the chlorite radical [129].

Applications of Sodium Peroxide. Sodium peroxide is produced in a number of countries. For example, in 1952 two companies in the United States, DuPont and United States Industrial Chemicals, produced approximately 4000 tons of Na_2O_2 containing not less than 19.7 wt.% of active oxygen corresponding to a product purity of 96% [57,130]. Another large American company, the National Distillers Products Corporation, also produces sodium peroxide. According to reference [54], in 1958 the United States produced approximately 8000 tons of sodium peroxide, consuming for this purpose approximately 5% of the sodium produced in the country. A product containing 96-98% Na_2O_2 is produced in Great Britain [131, 132] by the Imperial Chemical Industries, and in West Germany by the Degussa Company [56]. According to technical specification No. 1665-50 of the Soviet Ministry of the Chemical Industry, domestic sodium peroxide must contain not less than 94% Na_2O_2 [133]. In 1959, the selling price for 1 kg of Na_2O_2 was: 2 rubles in the United States, 2.15 rubles in France, 2.30 rubles in Italy, and 3.60 rubles in West Germany.

Basically, sodium peroxide finds application in bleaching of cotton, linen, wool fabrics, and jute materials, as well as rayon. References [131] and [132] describe various specifications to be followed for compounding bleaching solutions depending on the material to be bleached. The specifications were compiled taking into consideration the specific pH of the solution and stability of the hydrogen peroxide formed in the hydrolysis process of sodium peroxide. For this purpose, in most cases, specific quantities of sodium peroxide and stabilizers such as sodium silicate or sodium phosphate are added to the bleaching bath consisting of aqueous sulfuric acid or sodium carbonate. Sodium peroxide is also used for bleaching yarn or fabrics, either in a single-stage process, or after wash-out. It is also used in combination with other bleaching substances, for example, with sodium hypochlorite and sodium chloride. In bleaching jute materials in a single-stage process, sodium peroxide produces better color than hypochlorite. Moreover, the process is less expensive, since jute requires large quantities of chlorine, and the bleached fabric retains its strength [132].

Sodium peroxide is also used in bleaching wood products, ground wood, sulfate and sulfite cellulose, old paper products, semichemical products [133-143], viscous mass [144], straw [130-145], reeds [132], and many other materials, such as sponges [145], gelatin [132], etc.

In 1950, the bleaching of wood products with peroxides in the United States was used in 24 plants with daily productions of 1250 tons; in 1962, one-half of the sodium peroxide produced was used for such purpose. The bleaching of ground wood with sodium peroxide produces better results than with any other process [132]. Bleaching of ground sulfate or sulfite with chlorine is feasible only to a certain limit; above it, the strength of the product decreases. The use of sodium peroxide, instead of caustic soda, in the second stage of the alkaline extraction improves the color without any weakening of the product's strength [132]. The use of sodium peroxide for bleaching paper waste-products is convenient for removal of typographical colors, and in semichemical products it is advantageous because of low weight losses [132, 146].

The chemical reactions which take place in the bleaching of wood products with peroxides have not been studied sufficiently. Previous hypotheses stating that atomic oxygen participates in the process have been discounted. Today, it is generally accepted that HO_2^- ions are the active bleaching reagent [134]. In the bleaching process of wood products, sodium peroxide and hydrogen peroxide are mutually replaceable, with one kilogram of sodium peroxide being equivalent to 1.19 kg of 35% hydrogen peroxide solution. In order to maintain the required alkalinity (pH 10-10.5), caustic soda must be added to hydrogen peroxide bleaching solutions. With sodium peroxide solutions, the alkalinity is excessive and must be

neutralized with sulfuric acid. Usually, about 5% of sodium
metasilicate is added to the bleaching solution as a buffer, and
magnesium sulfate is added as a stabilizer.

Sodium peroxide is used as the starting material in the
syntheses of sodium superoxide [15], sodium metaborate perhydrate [55], sodium carbonate perhydrate [132], and other
peroxide compounds including organic peroxides [130] such
as benzoyl peroxide and lauryl peroxide. These peroxides
find wide application as polymerization initiators in the production of plastic products. They are synthesized by the reaction of the corresponding halogen compound with a strongly
basic peroxide solution. In this process sodium peroxide is
used since it is the most economical material.

Sodium peroxide has been extensively used as a component
in mixtures used for air regeneration in enclosed spaces
and in self-contained respiration apparatus [130, 147, 148]
since it reacts with moisture and carbon dioxide, liberating
oxygen and absorbing CO_2.

Alkaline solutions containing 50-100 g/liter of sodium
peroxide and 300-400 g/liter sodium oxide hydrate are used in
the treatment of metallic surfaces, especially in the bluing of
iron and steel [149]. Sodium oxide hydrate, containing 0.5-3%
sodium peroxide, is used for the removal of scale from stain-
less steel [150].

A mixture of sodium peroxide and sodium oxide hydrate
(4:1) is used in the decomposition of columbite to produce
tantalum and niobium salts [151], and also in the decomposition
of titanium magnetite ores and titanium concentrates [152].
Additions of sodium peroxide, mixtures of sodium peroxide
and hydrogen peroxide, or sodium peroxide and lithium fluo-
ride to an aqueous suspension of crushed phosphate ore simpli-
fies its enrichment by the anion foam flotation method [153].

The ability of sodium peroxide to react with refractory
oxides, sulfates, silicates, and certain alloys to form compounds which are soluble in water is used in analytic chemistry
[153a, 154]. This peroxide is also used in melting chromite
and other chromium ores [154a], talc, rhodonite, beryl, rutile,

garnierite [153a], platinum ores [154b], and ceramics based on BeO and UO_2 [154c]; in determining vanadium in ores, silicates, slags, and alloys [155]; and in determining iron in ferro—chromium and in iron—silicon alloys [154a]. In analytical chemistry, sodium peroxide is used for determining microquantities of phosphorus [156], for the colorimetric determination of cobalt and copper [157], and for the removal of iron from salt solutions [130]. Sodium peroxide is also used for decolorizing optical glass [158], and prevents the growth of algae in vessels designed for the extended storage of water. In organic synthesis, sodium peroxide can be used for the oxidation of mercaptans to form bisulfides. One patent [159] describes the use of sodium peroxide as a component of a solid rocket fuel consisting of a mixture of ammonium nitrate, 2,4—dinitrophenol, trinitrotoluene, collodium, and acrylic glue. Another patent [160] describes the application of an azide-sodium peroxide mixture (1.55 parts and 1 part, respectively) for transferring the rocket fuel and oxidizer from the rocket tanks to the combustion chamber.

Potassium Peroxide – K_2O_2

Potassium peroxide, which is quite unstable, being readily oxidized to KO_2 in air, can be obtained by passing a specific quantity of oxygen through a solution of metallic potassium in liquid ammonia at $-50°C$ [161]. The necessity of controlling the quantity of oxygen is dictated by the fact that K_2O_2 can be oxidized to KO_2 even in liquid ammonia.

Potassium peroxide is also formed in other reactions; for example, it can be produced by heating potassium oxide, K_2O, in a vacuum (10^{-5} mm Hg) at temperatures above $450°C$ [162], i.e.,

$$2K_2O(solid) \rightarrow K_2O_2(solid) + 2K(gas)$$

It can be obtained from potassium superoxide by reaction with chlorine dioxide diluted with carbon tetrachloride [163], or by the decomposition of the superoxide in a vacuum at 275-290°C

[41]. The latter method has been used for obtaining single crystals of K_2O_2 for x-ray structural analysis.

The density of K_2O_2 is 2.40 g/cm^3. It has a rhombic elementary cell with constants a = 6.736 A, b = 7.001 A, c = 6.479 A; Z = 4. Its x-ray density is 2.40 g/cm^3, molecular volume is 46.0 cm^3, and the space group is D_{2h}^{18} [41]. The K_2O_2 lattice energy is estimated at 482 kcal/mole [90]. Its refractive index is 1.456. The molecular refraction is 13.58 [164].

Reference [44] contains the following values for the $2K + O_2 \rightarrow K_2O_2$ reaction: $\Delta F°_{298}$ = −102 ± 11.0 kcal; $\Delta H°_{298}$ = −118.0 ± 10.0 kcal. Values of ΔF and ΔH up to 1500°K are also given.

The $K_2O_2 \cdot 2H_2O_2$ compound is known and is formed by the reaction of KOH with a concentrated solution of hydrogen peroxide. Upon decomposition, it is converted to KO_2 [118, 165, 166]. The crystal constants of $K_2O_2 \cdot 2H_2O_2$ are: a = 11.60 ± 0.05 A, b = 5.28 ± 0.03 A, c = 8.68 ± 0.05 A; $\beta \sim 75°$; the density is 2.30 g/cm^3; and the space group is C2/c No. 15 [115b].

Rubidium Peroxide — Rb_2O_2

Rubidium peroxide can be synthesized by a method similar to that used for potassium peroxide production [41, 162].

For the $2Rb + O_2 \rightarrow Rb_2O_2$ reaction, $\Delta H°_{298}$ = −116.5 ± 15.0 kcal; $\Delta F°_{298}$ = −95.5 ± 16.5 kcal. ΔH and ΔF values have been calculated for temperatures up to 1500°K [44]. The thermal decomposition of rubidium peroxide has been studied in the temperature range from 300 to 360°C and the thermodynamic constants of the process have been calculated [167].

The density of rubidium peroxide is 3.65 g/cm^3. The elementary cell is rhombic with constants a = 4.201 A, b = 7.075 A, c = 5.983 A; Z = 2. X-ray density is 3.802 g/cm^3. The molecular volume is 53.5 cm^3 and the space group is D_{2h}^{25} [41]. The lattice energy of rubidium peroxide is estimated as 464 kcal/mole [90].

Cesium Peroxide — Cs_2O_2

Cesium peroxide, like potassium and rubidium peroxides, can be synthesized by either the oxidation of metallic cesium in liquid ammonia with a calculated amount of oxygen, or more simply by the oxidation of metallic cesium with a calculated amount of oxygen. A specific amount of oxygen is necessary in both cases, since an insufficient amount of oxygen results in the formation of a series of suboxides with the composition Cs_7O to Cs_3O [168], along with Cs_2O, while an excess of oxygen yields either a mixture of the peroxide and superoxide, or the pure superoxide CsO_2.

The oxidation of an alkali metal—liquid ammonia solution was introduced by A. Joannis [169], who used it for the synthesis of potassium peroxide, K_2O_2. In the early 1900's, Rengade reported the synthesis of a cesium—oxygen compound as a result of the rapid oxidation of a liquid ammonia—cesium solution at −50 to −70°C. On the basis of the active oxygen content of the compound, 5.60%, it was assigned the formula Cs_2O_2 for which the theoretical active oxygen content is 5.43% [170-172]. In 1929, Kuhbier reported that the oxidation of metallic cesium in an aluminum vessel resulted in the formation of Cs_2O_2; however, no supporting analytical data were presented in the article [173].

Cesium peroxide has a yellowish color. According to Rengade [172], its melting point lies between 400-450°C, and in the molten state it does not decompose below 650°C. Its density, determined in toluene at 15°C, is 4.47 g/cm³.

Reference [44] includes the following thermochemical data for the reaction:

$$2Cs + O_2 \rightarrow Cs_2O_2$$

$\Delta H°_{298} = -96 \pm 15$ kcal; $\Delta F°_{298} = -79.5 \pm 17$ kcal; and the melting point of cesium peroxide is given as 597°C. The thermal decomposition of this compound has been studied in the temperature range from 300-340°C, and thermodynamic constants have been determined for the process [172a].

The crystal structure of cesium peroxide was not established until 50 years after its discovery. In 1957, H. Föppl [41] synthesized cesium peroxide by oxidizing metallic cesium dissolved in liquid ammonia. He obtained a compound with the empirical formula $Cs_2O_{2.2}$, and determined its lattice parameters by the powder method. The basic lattice of cesium peroxide is rhombic with parameters a = 4.322 A, b = 7.517 A, c = 6.430 A. The X-ray density, 4.74 g/cm^3, differs significantly from the value obtained by Rengade, who used a pycnometric method to obtain a value of 4.47 g/cm^3. The number of molecules in the basic lattice is two. The molecular volume is 63.0 cm^3, and the space group is D_{2h}^{25}. The lattice energy is estimated to be 445 kcal/mole [90].

The formation of cesium peroxide has also been reported as a result of the disproportionation reaction of cesium oxide in a vacuum (10^{-5} mm Hg) at 500°C [162], i.e.,

$$2Cs_2O \rightarrow Cs_2O_2 + 2Cs$$

Cesium peroxide peroxyhydrates, $Cs_2O_2 \cdot xH_2O_2$, are known. These materials have been obtained by the reaction of a cesium oxide hydrate solution with hydrogen peroxide at temperatures below 0°C [173]. The compounds have low stability and decompose above 1°C. Certain authors [116] classify these compounds as hydroperoxide peroxyhydrates having the formula $CsOOH \cdot H_2O_2$.

Ammonium Peroxide – $(NH_4)_2O_2$

It was believed that ammonium peroxide, $(NH_4)_2O_2$, was formed by the reaction of ammonia with hydrogen peroxide at −40°C [174]. Detailed studies of the $NH_3-H_2O_2$ system [175] revealed the presence of two compounds: $2NH_3 \cdot H_2O_2$ and $NH_3 \cdot H_2O_2$. The $2NH_3 \cdot H_2O_2$ compound melts partially at −93.5°C, and with the removal of ammonia, changes into the $NH_3 \cdot H_2O_2$ compound which melts at 22-23°C. The melting of $NH_3 \cdot H_2O_2$ results in an irreversible exothermic decomposition which terminates at +45°C with the liberation of oxygen [176].

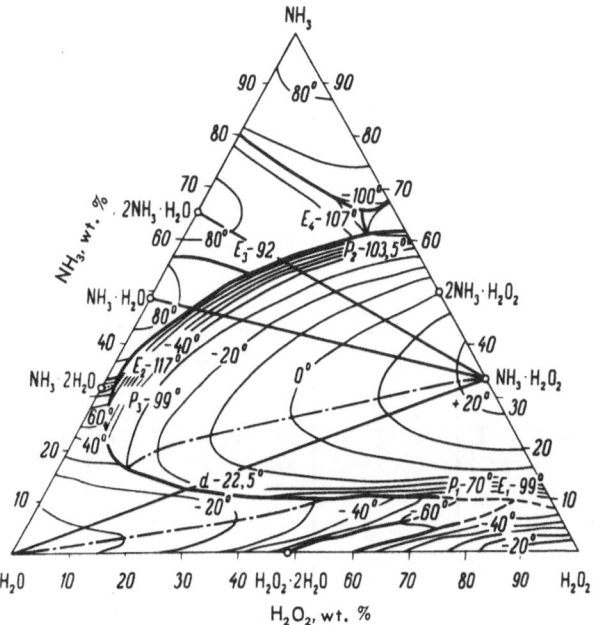

Fig. 6. Phase diagram of the $NH_3-H_2O_2-H_2O$ system [176].

A study of the $NH_3-H_2O_2-H_2O$ system [176] revealed that the characteristic feature of this system is the presence of five binary compounds: $NH_3 \cdot H_2O_2$, $2NH_3 \cdot H_2O_2$, $NH_3 \cdot 2H_2O$, $2NH_3 \cdot H_2O$, and $H_2O_2 \cdot 2H_2O$ (Fig. 6).

Giguère [177] has reported, based on his studies of the $NH_3-H_2O_2-H_2O$ system, the existence of the ternary compounds $NH_3 \cdot 3H_2O_2 \cdot 6H_2O$ melting at $-28.9°C$ and $(NH_4)_2O_2 \cdot 2H_2O_2$, which may be represented as $2NH_3 \cdot H_2O_2 \cdot 2H_2O$, melting at approximately $+15°C$. However, Mironov [178, 179] has shown that a detailed analysis of the phase diagram for the $NH_3-H_2O_2-H_2O$ system does not support an interpretation favoring the existence of ternary compounds. Raman [180] and infrared [181] spectra indicate that in the liquid state $NH_3 \cdot H_2O_2$ is a molecular compound, whereas in the solid state it undergoes

TABLE III

Basic Properties of Alkali Metal Peroxides

Formula	Active oxygen content, wt. %	Density, g/cm³	Melting point, °C	Disintegration temp., °C	ΔH°_{298}, kcal/mole	ΔF°_{298}, kcal/mole	ΔS°_{298}	ΔH°_{298} (from oxide and oxygen), kcal/mole	Lattice	U, kcal/mole
Li_2O_2	34.8	2.363	(425)	315–342	-152 ± 4.0	-136	(-26)	-9.6 ± 3	hexagonal	(578)
Na_2O_2	20.5	2.60	> 596	(636)	-122 ± 1.2	-112	-25.4	-22.6 ± 1.3	"	(532)
K_2O_2	14.5	2.40	(490)	—	-118 ± 10	-102 ± 11.0	—	-31.6 ± 6	rhombic	(482)
Rb_2O_2	7.9	3.65	(567)	—	-101.5 ± 10	-86.5 ± 11.5	—	-37.5 ± 9.5	"	(464)
Cs_2O_2	5.43	4.47	(597)	—	-96 ± 15	-79.5 ± 17.0	—	(-40)	"	(445)

Note:　Estimated values are in parentheses.

transformation to the hydroperoxide form NH_4OOH, i.e.,

$$NH_3 \cdot H_2O_2 \rightarrow NH_4^+ + HO_2^-$$

Raman spectra of ammonia–hydrogen peroxide solutions do not indicate the presence of the HO_2^- ion, whereas the Raman spectra of solutions of lithium, sodium, and potassium hydroxides in hydrogen peroxide do indicate the existence of the HO_2^- ion [182].

Some of the basic properties of the alkali metal peroxides are given in Table III. From the data presented in Table III, it can be seen that as the atomic number of alkali metal increases, the absolute value for the heat of formation of the peroxide decreases. According to Kapustinskii's thermodynamic logarithmic rule, a plot of the ratio of $\Delta H^\circ{}_{298}/W$ (where W is the valence of the metallic element) versus the log Z should show a straight-line relationship. This relationship is shown in Fig. 7.

The heats of formation of the peroxides were determined by means of the following relationship:

$$-\Delta H = -\lambda_m - I_m + E_{O_2^{2-}} + U_{m_2O_2}$$

where λ_m is the heat of vaporization of the metal, $E_{O_2^{2-}}$ is the energy associated with the formation of the O_2^{2-} ion from the O_2 molecule, I_m is the ionization potential of the metal, and

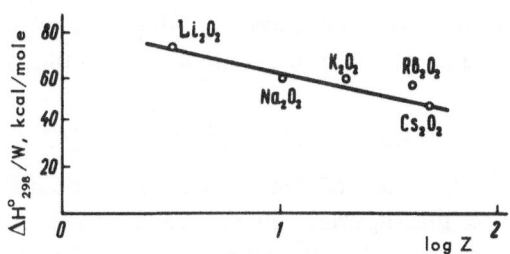

Fig. 7. $\Delta H^\circ{}_{298}/W$ ratio as a function of log Z for alkali metal peroxides.

TABLE IV

Heat of Evaporation, λ_m, and Ionization
Energy, I_m, of Alkali Metals*

Metal	λ_m, kcal/mole	I_m, kcal/mole
Li	37.07	162.86
Na	25.98	146.01
K	21.51	123.07
Rb	20.51	118.29
Cs	18.83	110.08

*Reference [183].

$U_{m_2O_2}$ is the lattice energy. As seen in Table IV, the absolute values of λ_m and I_m decrease as the atomic number of the elements increase. Estimated values for the lattice energies [90] (Table III) decreases more rapidly and consequently the heat of formation of the peroxides is smaller as one goes from Li_2O_2 to Cs_2O_2.

The ability of the alkali metals to form peroxides can be explained as follows. The lattice energy of alkali metal oxides is lower than the lattice energy of other metal oxides and, consequently, transformation to the peroxide lattice is possible. The peroxide lattice energy is less than the energy of the corresponding oxides. The energy required to form peroxides is furnished by the energy of formation of O_2^{2-} ions from O^{2-} ions and oxygen.

According to S. A. Shchukarev [184], the transition of oxides to peroxides takes place as a result of the exothermal process

$$M_2O^{2-} + O = M_2(O-O)^{2-}$$

which results because of the formation of coordinate bonds between the oxygen atoms. In the case of lithium oxide, the O^{2-} ion is stabilized by the small cation. To obtain lithium peroxide from the oxide and oxygen, it would be necessary to apply a very high oxygen pressure. As has been mentioned previously, lithium peroxide is formed by the reaction of lithium hydroxide with hydrogen peroxide in a way similar to the formation of an insoluble salt.

REFERENCES

1. P. George. J. Chem. Soc. (1955), p. 2367.
2. J. Bennet. Phil. Mag. 46:443 (1955).
2a. W. J. Sax. Dangerous Properties of Industrial Materials, New York, Reinhold Publishing Corp. (1962), p. 1127.
3. A. J. Cohen and J. Margrave. Anal. Chem. 29:1462 (1957).
4. J. Kleinberg. Unfamiliar Oxidation States and Their Stabilization, Lawrence, Kansas (1950).
5. W. S. Graff. J. Electrochem. Soc. 105:446 (1958).
6. P. Ray and D. Sen. Chemistry of Bi- and Tripositive Silver, Natl. Inst. Sci. India (1960).
7. A. Glassner. J. Chem. Soc. (1951), p. 904.
8. F. T. Magg and D. Sutton. Trans. Faraday Soc. 54:1861 (1958).
9. F. T. Magg and D. Sutton. Trans. Faraday Soc. 55:974 (1959).
10. S. Z. Makarov, T. I. Arnold, N. N. Stasevich, and E. V. Shorina. Izv. Akad. Nauk SSSR, Otd. Khim. Nauk (1960), p. 1913.
11. V. Frei. Collection Czech. Chem. Commun. 27:179 (1962).
12. V. Frei. Collection Czech. Chem. Commun. 27:430 (1962).
13. S. Z. Makarov and T. I. Arnold. Izv. Akad. Nauk SSSR, Otd. Khim. Nauk (1960), p. 2090.
14. A. B. Neiding and I. A. Kazarnovskii. Dokl. Akad. Nauk SSSR 78:713 (1951).
15. Inorganic Syntheses, Vol. 4, New York, McGraw-Hill Book Company (1953), p. 12.
16. J. A. MacMillan. Acta Cryst. 7:640 (1954).
17. G. M. Schwab and G. Hartmann. Z. Anorg. Allgem. Chem. 281:183 (1955).
18. T. Palagyi and I. Náray-Szabó. Acta Chim. Acad. Sci. Hung. 30:1 (1962).
19. V. A. Lunenok-Burmakina and A. I. Brodskii. Dokl. Akad. Nauk SSSR 129:1335 (1959).
20. J. A. MacMillan. J. Inorg. Nucl. Chem. 13:28 (1960).
21. J. A. MacMillan. Chem. Rev. 62:65 (1962).
22. R. de Forcrand. Compt. Rend. 130:1465 (1900).
23. F. Feher. Ber. 86:1429 (1953).
24. S. Z. Makarov and T. A. Dobrynina. Izv. Akad. Nauk SSSR, Otd. Khim. Nauk (1955), p. 411.
25. T. A. Dobrynina. Author's Candidate Thesis, Moscow, IONHKh Akad. Nauk SSSR (1957).
26. T. A. Dobrynina. Izv. Akad. Nauk SSSR, Otd. Khim. Nauk (1960), p. 962.
27. S. Z. Makarov and T. A. Dobrynina. Izv. Akad. Nauk SSSR, Otd. Khim. Nauk (1956), p. 294.
27a. T. A. Dobrynina. Lithium Peroxide, Moscow, Izd. "Nauka" (1964).
28. A. J. Cohen. J. Am. Chem. Soc. 74:3762 (1952).
29. J. Aubry and Ch. Gleitzer. Bull. Soc. Chim. France (1957), p. 109.
30. H. H. Strater. U. S. Patent 2292358 (1960).
31. R. O. Bach and W. W. Boardman. Chem. Eng. News 40(47):54 (1962).

32. H. Hahn. Z. Anorg. Allgem. Chem. 275:35 (1954).
33. D. Schechter and J. Kleinberg. J. Am. Chem. Soc. 76:3297 (1954).
34. H. Lux. Z. Anorg. Allgem. Chem. 298:298 (1959).
35. S. Z. Makarov and T. A. Dobrynina. Izv. Akad. Nauk SSSR, Otd. Khim. Nauk (1960), p. 1321.
36. I. I. Vol'nov. Izv. Akad. Nauk SSSR, Otd. Khim. Nauk (1957), p. 762.
37. I. I. Vol'nov and A. N. Shatunina. Zh. Neorgan. Khim. 4:257 (1959).
38. T. V. Rode, T. A. Dobrynina, and G. A. Gol'der. Izv. Akad. Nauk SSSR, Otd. Khim. Nauk (1955), p. 61.
39. K. Notz and R. Bach. Chimia (Aarau) 17:158 (1963).
40. M. Markowitz, D. A. Boryta, and H. Stewart. Chem. Eng. News 41(3):5 (1963).
41. H. Föppl. Z. Anorg. Allgem. Chem. 291:46 (1957).
42. E. Brame. J. Inorg. Nucl. Chem. 4:90 (1957).
43. L. Brewer. Chem. Rev. 52:6 (1953).
44. J. P. Coughlin. Bull. 542 Bureau of Mines, Washington (1954).
45. K. I. Selezneva. Zh. Neorgan. Khim. 5:1688 (1960).
46. J. H. Lanmeck. J. Chem. Eng. Data 6:233 (1960).
47. W. D. Johnston and R. R. Heikes. J. Chem. Phys. 26:582 (1957).
48. W. D. Johnston and R. R. Heikes. J. Phys. Chem. 7:1 (1958).
49. W. D. Johnston and R. R. Heikes. J. Am. Chem. Soc. 78:3255 (1956).
50. R. R. Heikes and W. D. Johnston. U. S. Patent 2921973 (1960).
50a. R. R. Heikes and W. D. Johnston. U. S. Patent 2953617 (1960).
51. W. D. Johnston, R. C. Miller, and R. Maselsky. J. Phys. Chem. 63:198 (1959).
52. C. Walling and D. A. Buckler. J. Am. Chem. Soc. 77:6032 (1955).
53. H. J. Castner. German Patent 67094 (1891).
54. M. Sittig. Sodium, Moscow, Gosatomizdat (1961), p. 192 [originally published in English by Reinhold Publishing Corp., New York].
55. D. W. Hardie. Ind. Chemist 31:385 (1955).
56. J. Miller. Chem. Ind. 2:91 (1953).
57. H. R. Tennant and R. B. Schow. Handling and Use of the Alkali Metals, Advan. Chem. Ser. 19, ACS, New York (1957), p. 118.
58. R. Kohlmuller. Ann. Chim. 4:1190 (1959).
59. D. S. Natz. U. S. Patent 2633406 (1953).
60. R. F. Hulse and D. S. Natz. U. S. Patent 2685500 (1954).
61. National Distillers Prod. Corp. British Patent 730130 (1955).
62. Chem. Eng. News 34:1992 (1956).
63. W. Klabunde and J. Joung. U. S. Patent 2752226 (1956).
64. Chem. Trade J. 138(3598):1198 (1956).
65. T. Tadler and R. Coleman. U. S. Patent 2789885 (1957).
66. F. Schumacher and G. Irwin. U. S. Patent 2825629 (1958).
67. L. I. Governale. U. S. Patent 2671010 (1954).
68. Chem. Eng. 69(6):90 (1962); J. Szechtman. U. S. Patent 3119664 (1964).
69. R. Schow and R. Coleman. C. A. 51:13729 (1957).
70. F. T. Seelye and T. A. Rafter. Nature 165:316 (1950).
71. G. J. Petretic. Anal. Chem. 23:1183 (1951).
72. R. P. Anibal. Anal. Chem. 32:293 (1960).
73. R. S. Joung and K. G. A. Strachan. Chem. Ind. 7:154 (1953).
74. O. Rosner. Z. Metallk. 48:137 (1958).
75. H. E. Blake and W. F. Holbrook. Chemist–Analyst 46:42 (1957).
76. R. L. Tallman. J. Am. Chem. Soc. 79:2979 (1957).
77. H. Hohn. Austrian Patent 175237 (1953).
78. G. Jangg. Z. Anorg. Allgem. Chem. 311:186 (1961).
79. H. W. Nicolai. Chim. Ind. 73:1156 (1955).
80. H. Ostertag and J. Chassain. Compt. Rend. 238:684 (1954).

81. A. Etienne and J. Fellion. Compt. Rend. 238:1429 (1954).
82. H. Hock, H. Knopf, and E. Ernst. Angew. Chem. 71:541 (1959).
83. B. Emmert. Ber. 54B:204 (1921).
84. R. Setton. U.S. Patent 2083691 (1958).
85. R.M.R. Ornhjieim. Austrian Patent 202974 (1959); West German Patent 1054970 (1959).
86. S.Z. Makarov and N.K. Grigor'eva. Izv. Akad. Nauk SSSR, Otd. Khim. Nauk (1955), p. 17.
87. S.Z. Makarov and N.K. Grigor'eva. Izv. Akad. Nauk SSSR, Otd. Khim. Nauk (1955), p. 208.
88. R.L. Tallman. Dissertation Abstr. 20:4293 (1960).
89. R.L. Tallman and J.L. Margrave. J. Inorg. Nucl. Chem. 21:40 (1961).
90. K.B. Yatsimirskii. Izv. VUZOV, Khim. i Khim. Tekhnol. 2:480 (1959).
91. C. Duval and J. Lecomte. Bull. Soc. Chim. France 20(5):205 (1953).
92. T.R. Griffiths, K.A. Lott, and M.C.R. Symons. Anal. Chem. 31:1338 (1959).
93. P.W. Gilles and J.L. Margrave. J. Phys. Chem. 60:1333 (1956).
94. S.S. Todd. J. Am. Chem. Soc. 75:1229 (1953).
95. M.S. Chandrasckaraiah. J. Phys. Chem. 63:1505 (1959).
96. K. Savithri and S.R. Rao. Proc. Indian Acad. Sci. 16A:221 (1942).
97. T.V. Rode and G.A. Gol'der. Izv. Akad. Nauk SSSR, Otd. Khim. Nauk (1956), p. 299.
98. T.V. Rode and G.A. Gol'der. Dokl. Akad. Nauk SSSR 110:1001 (1956).
99. M. Jacquinot. Compt. Rend. 238:105 (1954).
100. M. Jacquinot. Compt. Rend. 239:61 (1954).
101. M. Viltange. Ann. Chim. 5:1037 (1960).
102. M. Viltange. Chim. Anal. 42:608 (1960).
103. M. Viltange-Jacquinot. Compt. Rend. 242:781 (1956).
104. M. Viltange-Jacquinot. Compt. Rend. 244:1215 (1957).
105. C. Duval and J. Lecomte. Compt. Rend. 234:2445 (1952).
106. Chem. Eng. News 32:258 (1954).
106a. Ya. I. Mikhailenko. Zh. Russ. Fiz.-Khim. Obshchest. 53:350 (1921).
107. A.A. Zalutraeva. Tr. Gos. Inst. Prikl. Khim. No. 45:97 (1960).
108. J. Mattner and R. Mattner. Z. Anal. Chem. 134:524 (1951).
109. Inorganic Syntheses, Compl. 3, Moscow, IL (1952), p. 7.
110. E. Nast. Angew. Chem. 65:266 (1953).
111. R. Nast. Oesterr. Chemiker-Ztg. (1953), p. 152.
112. Structure Reports 8:125 (1956).
113. S. Linderborg. Kymi Ytyma No. 1:3 (1960).
114. O. von Schink. Chem. Ingr.-Tech. 37:462 (1960).
115. A. Kh. Mel'nikov and T.P. Firsova. Zh. Neorgan. Khim. 6:2470 (1961).
115a. P. Jaubert. Compt. Rend. 132:86 (1901).
115b. N.G. Vannerberg. In: Progress in Inorganic Chemistry, New York, Interscience Publishers, Inc. (1962), p. 173.
116. J. Partington and A. Fathallah. J. Chem. Soc. (1959), p. 1934.
117. I.I. Vol'nov and A.N. Shatunina. Zh. Neorgan. Khim. 4:1491 (1959).
118. G.L. Cunningham and F.R. Romesberg. U.S. Patent 2908552 (1959).
119. Chem. Eng. News 31(39):4012 (1953).
120. A. Le Berre. Compt. Rend. 252:1341 (1961); French Patent 1290179 (1962).
121. A. Le Berre. Bull. Soc. Chim. France (1961), p. 1198.
122. O. Glemser, H. Hanschild, and G. Lutz. Z. Anorg. Allegem. Chem. 269:93 (1952).
123. T.V. Rode, A.P. Zachat-skaya, and G.A. Gol'der. Zh. Neorgan. Khim. 5:524 (1960).
124. C.A. Addison and J. Lewis. J. Chem. Soc. (1953), p. 1869.

125. K.E. Mironov, B.S. Dzyatkevich, and T.I. Rogozhnikova. Izv. Sibirsk. Otd. Akad. Nauk SSSR 11:130 (1962).
126. J. Rademachers and U. Wannagat. Angew. Chem. 69:783 (1957).
127. M. Schmidt and H. Bipp. Z. Anorg. Allgem. Chem. 303:205 (1960).
128. M. Antelman. J. Chem. Educ. 32:273 (1955).
129. L. Duval. C.A. 48:11926 (1954).
130. R. Kirk and D. Othmer. Encyclopedia of Chemical Technology, Vol. 10, New York, The Interscience Encyclopedia (1953), p. 38.
131. A.S. White. Dyer 111:417 (1954).
132. Imperial Chemical Industries Limited. Granular Sodium Peroxide, Birmingham, Kynoch Press (1962), p. 25.
133. Yu. V. Karyakin and I.I. Angelov. Pure Chemical Reagents, Moscow, Goskhimizdat (1955).
134. L.A. Biman et al. Bleaching of Cellulose, Moscow–Leningrad, Goslesbumizdat (1957), p. 201.
135. C.A. 48:10341, 11055, 13219 (1954).
136. A. Yankovskii. Khim. i Khim. Tekhnol. 10:134 (1956).
137. G. Rowlendson. Khim. i Khim. Tekhnol. 10:146 (1956).
138. F. Wultsch. Tappi 42:313 (1959).
139. W.F. Schroeder. U.S. Patent 2865701 (1958).
140. W. Gertner. West German Patent 873651 (1953).
141. M.J. Mouton, J.M. Cholley, A. Meunier–Guttin. Papeterie 76:311, 357; 77:339 (1955); 79:205 (1957).
142. G.L. Bergada and S. Hernandes. Rev. Cienc. Apl. 7:418 (1953).
143. D. Berruso and G. Ceragioli. Ind. Carta 8:11 (1954).
144. H. Sihtola. Paperi Puu 34:447 (1957).
145. W. Hundt and K. Wieweg. Seifen-Oele-Fette-Wachse 81:419, 444 (1955).
146. P. Kajanns and V. Ostring. Paperi Puu 40:203 (1958).
147. F.E. Clarke. J. Am. Soc. Naval Engrs. 68:105 (1956).
148. Auergesellschaft. West German Patent 1009929 (1957); RZh. Khim. No. 3:8823 (1959).
149. P.H. Margulis. Plating 42:561 (1955).
150. C.B. Francis. U.S. Patent 2569159 (1951).
151. C.A. 48:7857 (1954).
152. N.R. Shvedov. Zavodskaya Lab. 20:915 (1954).
153. J. Le Baron. U.S. Patent 2826301 (1958).
153a. T. Rafter. Analyst 75:485 (1950).
154. K.F.G. Hosking. Mining Mag. 89:137 (1953).
154a. C.B. Belcher. Talanta 10:75 (1963).
154b. T.R. Cunningham and T.R. McNeil. Ind. Eng. Chem. (Analyt. Edition) 1:70 (1929).
154c. F.M. Postma, Jr. U.S. At. Energy Comm. Y:1377 (1962).
155. Analyst 80:391 (1955).
156. J. Saje. Kohászati Lapok 14:383 (1959).
157. R. Rabea. Analele Stiint. Univ. "A.I. Cuza" Iasi, Sect. I (N.S.) 4:171 (1958).
158. C. Weissenberg and N. Meinert. U.S. Patent 2763559 (1956).
159. F. Salvini. Italian Patent 516030 (1954).
160. U.S. NASA. U.S. Patent 2981616 (1961).
161. W. Schumb, et al. Hydrogen Peroxide, Moscow, IL (1958) [originally published in English by Reinhold Publishing Corp., New York].
162. W. Klemm and H.J. Scharf. Z. Anorg. Allgem. Chem. 303:263 (1960).
163. C. Bertoglio Riolo. Ann. Chim. 44:815 (1954).
164. I.A. Kazarnovskii and S.I. Raikhshtein. Zh. Fiz. Khim. 21:252 (1947).
165. I.A. Kazarnovskii and A.B. Neiding. Dokl. Akad. Nauk SSSR 86:717 (1952).

166. A. W. Petrocelli and D. L. Kraus. J. Chem. Educ. 40:146 (1963).
167. A. W. Petrocelli. Dissertation Abstr. 21:1081 (1960); J. Phys. Chem. 66:1225 (1962).
168. G. Brauer. Z. Anorg. Chem. 255:101 (1947).
169. A. Joannis. Compt. Rend. 116:1370 (1893).
170. E. Rengade. Compt. Rend. 140:1536 (1905).
171. E. Rengade. Bull. Soc. Chim. (Paris) 35(3):769 (1906).
172. E. Rengade. Ann. Chim. Phys. 11(8):348 (1907).
172a. G. V. Morris. Dissertation Abstr. 23:2343 (1963).
173. F. Kuhbier. Dissertation, Berlin Friedrich-Wilhelms Universität (1929).
174. Hydrogen Peroxide and Peroxide Compounds, edited by M. E. Pozin, Moscow, Goskhimizdat (1951).
175. K. E. Mironov. Zh. Neorgan. Khim. 4:153 (1959).
176. K. E. Mironov. Candidate Thesis, Moscow, IONKh Akad. Nauk SSSR (1953).
177. P. Giguère and D. Chin. Can. J. Chem. 37:2064 (1959).
178. K. E. Mironov. Izv. Sibirisk. Otd. Akad. Nauk SSSR (1960), p. 143.
179. K. E. Mironov. Can. J. Chem. 38:2269 (1960).
180. A. Simon and K. Krishman. Naturwiss. 42:14 (1955).
181. A. Knop and P. Giguère. Can. J. Chem. 37:1794 (1959).
182. A. Simon and M. Marchand. Z. Anorg. Allgem. Chem. 262:192 (1960).
183. F. D. Rossini. Selected Values of Chemical Thermodynamic Properties, Circular 500, NBS US, Washington (1952).
184. S. A. Shchukarev. Zh. Obshch. Khim. 28:857 (1958).

Chapter Three

Peroxides of the Group Two Elements of the Periodic Table

All elements of the second group, with the exception of beryllium [1], form peroxide compounds. Peroxides of calcium, strontium, and barium belong to the $M^{2+}O_2^{2-}$ type; while peroxides of magnesium, zinc, and cadmium, of general formula $MO_2 \cdot xH_2O$, probably belong to the HO—M—OOH type where the covalent bond between the hydroperoxyl group and the metal atom is the same as that in the hydrogen peroxide molecule [2-4]. Radium peroxide, RaO_2, the heat of formation of which is estimated as 150 kcal/mole [5], has not yet been produced.

Unlike sodium peroxide and the peroxides of the potassium subgroup metals, peroxides of the group two elements cannot be synthesized by direct oxidation of the metals. Barium peroxide, BaO_2, can be obtained by oxidizing BaO; peroxide compounds of the other elements of this group are obtained by reacting the corresponding hydroxides with hydrogen peroxide solution. All of these methods are well known. Investigations conducted in recent years have been dedicated mainly to the study of the properties and structure of the peroxides and their hydrates and peroxyhydrates.

Of the aforementioned peroxides, barium peroxide, in the main, and to a lesser degree, calcium peroxide, CaO_2, and the hydrate forms of magnesium and zinc peroxides have practical applications. The applications of strontium peroxide, SrO_2, are quite limited, whereas mercury and cadmium peroxides have no practical uses. In pure form, the peroxides of calcium, strontium, and barium and the hydrated forms of magnesium, zinc, and cadmium peroxides are colorless and diamagnetic. Mercury peroxide, HgO_2, is yellow, while industrial grade CaO_2, SrO_2, and BaO_2 have a cream color which is probably caused by the presence of superoxide [6]. The peroxides of calcium, strontium, and barium are more stable in the presence of moisture and carbon dioxide than the peroxides of the alkali metals. They cannot be decomposed by water, nor do they dissolve in water or organic substances. They react with diluted aqueous acid solutions to form the corresponding salts and hydrogen peroxide. Determination of the active oxygen content of the alkaline earth peroxides with standard permanganate is performed in an acidic medium, using orthophosphoric acid rather than sulfuric acid, to prevent the formation of insoluble sulfates.

ALKALINE EARTH PEROXIDES

Calcium Peroxide — CaO_2

Production Methods. Calcium peroxide can be obtained by the dehydration of $CaO_2 \cdot 8H_2O$. The latter is synthesized by the reaction of dilute hydrogen peroxide with calcium oxide hydrate suspensions, or by the reaction of calcium salt solutions (chloride or nitrate) with ammonia and hydrogen peroxide [7-9]. In the United States [10] and Great Britain [11], the first method is used for the production of an industrial product which contains 75-60 wt.% of CaO_2. By using the second method, it is feasible to obtain products with higher purity in both the laboratory and under industrial conditions [12, 13]. To obtain $CaO_2 \cdot 8H_2O$ in high yield from calcium salts, aqueous ammonia solutions, and hydrogen peroxide, it is recommended that a 20% calcium chloride solution containing a $CaCl_2$:H_2O_2 ratio of 1:4 be poured into an aqueous mixture (cooled to 0-4°C) of hydrogen peroxide and ammonium oxide hydrate containing 2.5% H_2O_2 and 2% NH_3 [13].

Of significant value in the selection of the best conditions for the synthesis of $CaO_2 \cdot 8H_2O$ are the data obtained from a

TABLE V

Ca(OH)$_2$–H$_2$O$_2$–H$_2$O System

Solid phase	H$_2$O$_2$ content in liquid phase, wt.%				
	+50°C	+10°C	0°C	−10°C	−21°C
CaO$_2\cdot$8H$_2$O	—	0.0−0.10	0.0−6.0	14.8−21.8	24.5−28.0
CaO$_2\cdot$2H$_2$O	—	0.10−22.0	6.0−24.8	—	—
CaO$_2\cdot$0.5H$_2$O	1.7−47.5	—	—	—	—
CaO$_2\cdot$2H$_2$O$_2$	47.5−87.7	22.0−78.5	24.8−88.0	21.8−91.0	28.0−82.0

study of the Ca(OH)$_2$–H$_2$O$_2$–H$_2$O system at −21°C, −10°C [14], +10°C [15], and +50°C [16]. The thermal stability data on CaO$_2\cdot$8H$_2$O are also important for selecting the best conditions for its conversion to CaO$_2$ [17].

Table V shows the limits of the hydrogen peroxide equilibrium concentrations for solid phase regions of the Ca(OH)$_2$–H$_2$O$_2$–H$_2$O system at the above-mentioned temperatures.

The CaO content in the liquid phases of the system varies from 0.06 to 0.90 wt.% depending on the concentration of hydrogen peroxide, the temperature, and the solid phase. According to Table V, the system has four solid phases. However, the existence of CaO$_2\cdot$2H$_2$O and CaO$_2\cdot$0.5H$_2$O has been substantiated only by the "remainder" method.

To obtain a high yield of CaO$_2$ from the octahydrate, it is recommended that the dehydration be performed in two stages: first, CaO$_2\cdot$8H$_2$O is dehydrated by heating in water at 50°C; after filtration the deposit is vacuum dried at 100°C [17]. According to the heating curve (see Fig. 13), CaO$_2\cdot$8H$_2$O melts partially at 40-50°C. It undergoes dehydration at 100°C, and at 380°C the dehydrated CaO$_2$ decomposes to CaO +$\frac{1}{2}$O$_2$ [3, 17].

Another method for obtaining calcium peroxide which has not found practical application is the oxidation of calcium amalgam [18]. The reaction of calcium nitrate with KO$_2$ or NaO$_2$ in liquid ammonia at −30°C [19] results in the formation of products containing not more than 60% CaO$_2$. Attempts to obtain calcium peroxide by the reaction of calcium oxide with oxygen under pressure were unsuccessful. Investigation of the $CaO-O_2-CaO_2$ system revealed [20] that at 51.6 atm pressure and at 100°C, CaO and O_2 formed only traces of calcium peroxide (0.15%). The equilibrium pressure of oxygen in the above-mentioned system is approximately 70.3 atm at 105.5°C, and 107.2 atm at 110°C.

Properties of Calcium Peroxide.   Calcium peroxide has significant thermal stability. Heating it at atmospheric pressure in the temperature range from 175 to 300°C does not result in any meaningful liberation of active oxygen. Its intensive decomposition begins at 375°C and it disintegrates completely at 400-425°C [17].

Studies have been made of the dissociation kinetics in the temperature range from 300 to 350°C and at an oxygen pressure 10-760 mm Hg [21, 22]. The apparent energy of activation for the dissociation process is estimated at 45,000 kcal/mole. The variation of equilibrium pressure in the thermal decomposition of calcium peroxide indicates the formation of solid solutions with limited solubility.

The decomposition of calcium peroxide in the temperature range from 305 to 370°C under a high vacuum [23, 24] proceeds with the formation of solid solutions of peroxide and oxide. In a high vacuum, $CaO_2 \cdot 8H_2O$ dissociates at a faster rate and at a lower temperature (150-200°C) than CaO_2 [23, 24]. The authors of this work [23, 24] believe that this significant drop in the temperature of dissociation results from the formation of active calcium peroxide by the separation of the water of hydration and consequent lattice destruction. However, according to the data in reference [25], the dehydration process of $CaO_2 \cdot 8H_2O$ is accompanied by disproportionation of the O_2^{2-} ion to O_2^- and O^{2-} ions. It is also possible that there is formed in this process $Ca(O_2)_2$, which is more reactive with water vapor than is CaO_2 [26].

The density of CaO_2 is 2.92 g/cm^3 [27]. The unit cell is tetragonal, the same as hydrogen peroxide, and the cell constants are: a = 3.55 A, c = 5.98 A, c/a = 1.68; Z = 2. X-ray density is 3.19 g/cm^3; the molecular volume is 22.6 cm^3; and the space group is D_{4h}^{17} [27].

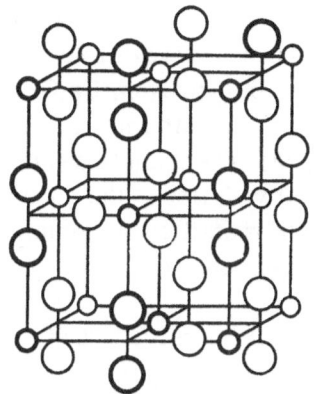

Fig. 8. Typical crystal lattice for peroxides of
alkaline earth metals [33].

Figure 8 shows a typical lattice for peroxides of the alkaline earth metals, where each metal ion (small circle) is surrounded by six peroxide ions (paired large circles) forming an octahedron.

For the reaction $\frac{1}{2}Ca + \frac{1}{2}O_2 \rightarrow \frac{1}{2}CaO_2$, $\Delta F°_{298} = -72$ kcal, $\Delta H°_{298} = -78.2 \pm 0.5$ kcal, $\Delta S°_{298} = -20$ entropy units (eu) [28]. The estimation of $\Delta S°_{298} = 10.3$ eu given in reference [29] is incorrect; the reported value is too low. Calcium peroxide is diamagnetic, $\chi_m = -23.8 \cdot 10^{-6}$ [30].

Although calcium perioxide has a higher active oxygen content than sodium peroxide (22.2 wt.% instead of 20.5%), it has more limited applications than the latter because of its insoluble nature and lower reactivity. Anhydrous calcium peroxide does not react with dry carbon dioxide at temperatures lower than the dissociation temperature even in the presence of catalysts [31]. In a stream of moist carbon dioxide, calcium peroxide begins to decompose at 145°C [31], although the reaction is noticeable at room temperature [26]. The decomposition rate can be increased severalfold by the addition of NaOH, MnO_2, and CuO [31]. Calcium peroxide octahydrate, $CaO_2 \cdot 8H_2O$, begins to decompose at 15°C in a stream of carbon dioxide without the presence of catalysts.

For mixtures of sodium oxide hydrate and MnO_2, the decomposition rate is increased severalfold [31]. In the reduction of calcium peroxide with hydrogen, $Ca(OH)_2$ is formed at lower temperatures (to 300°C). At higher temperatures, CaO_2 is reduced to CaO [21, 22]. The reduction rate of calcium peroxide with hydrogen initially follows the monomolecular reaction equation [23]. With metallic aluminum, calcium peroxide reacts via the following equations:

$$3CaO_2 + 2Al \rightarrow Al_2O_3 \cdot 3CaO + 385 \, kcal$$

$$2CaO_2 + 2Al \rightarrow Ca + CaO \cdot Al_2O_3 + 273 \, kcal$$

This process is used to advantage as an additional source of heat in certain metallurgical processes [32].

According to Table V, calcium peroxide can bind water of crystallization and hydrogen peroxide to form $CaO_2 \cdot 8H_2O$ and $CaO_2 \cdot 2H_2O_2$.

$CaO_2 \cdot 8H_2O$ crystallizes in the tetragonal system. The lattice constants are: a = 6.21 ± 0.03 A, c = 11.02 ± 0.05 A; Z = 2. Its density is 1.672 g/cm^3 [12]. Figure 9 shows a microphotograph (×475) of a $CaO_2 \cdot 8H_2O$ crystal and is typical for octahydrates of alkaline earth metal peroxides. The molecular structure of calcium peroxide octahydrate, as well as peroxide octahydrates of strontium and barium, can be represented as chains of $-O_2^{2-}-(H_2O)_8-O_2^{2-}$ rotating around the c-axis [33]. The water molecules and O_2^{2-} ions in the chains are connected by hydrogen bonds.

$CaO_2 \cdot 2H_2O_2$ is unstable, and at 50°C it liberates its bound hydrogen peroxide. $CaO_2 \cdot 2H_2O_2$ crystallizes in the monoclinic system in two modifications: α and β. The lattice parameters of the β-modification have not been established, while the lattice parameters of the α-modification are: a = 8.154 ± 0.005 A, b = 5.675 ± 0.003 A, c = 8.059 ± 0.005 A. The x-ray density is 2.553 g/cm^3, and the optical axis angle is 101°58' ± 4' [33]. The structure of calcium peroxide diperoxyhydrate, as well as of the diperoxyhydrates of strontium and barium peroxides, can be represented in the form of chains of

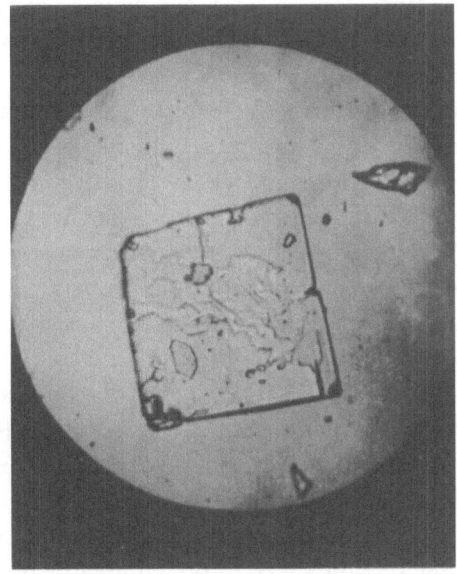

Fig. 9. $CaO_2 \cdot 8H_2O$ crystal ($\times 475$).

$O_2^{2-} \ldots (H_2O_2)_2 \ldots O_2^{2-} \ldots (H_2O_2)_2 \ldots$ [33], in which O_2^{2-} ions are connected to hydrogen peroxide molecules by strong hydrogen bonds.

The thermal decomposition of $CaO_2 \cdot 2H_2O_2$ can result in the formation of a mixture of CaO_2, $Ca(O_2)_2$, and $Ca(OH)_2$ provided that the process is allowed to take place in a vacuum at a residual pressure of less than 10 mm Hg, and below 50°C [34, 35, 36]. CaO_2 is formed when the decomposition takes place at atmospheric pressure and approximately 100°C, while CaO is formed when the reaction is carried out at atmospheric pressure and at temperatures above 375°C [17].

The reaction of $CaO_2^{18} \cdot 8H_2O$ with solutions containing greater than 20 wt.% of hydrogen peroxide results in the formation of the compound $CaO_2 \cdot 2H_2O_2$, in which all of the oxygen atoms of CaO_2 and of the bound hydrogen peroxide have similar isotopic composition, indicating complete isotopic oxygen exchange between calcium peroxide and hydrogen peroxide. At

H_2O_2 concentrations of 15 wt.% and below, oxygen exchange between calcium peroxide and hydrogen peroxide was not observed [37].

Applications of Calcium Peroxide. In the United States, calcium peroxide is used in bakeries to increase the plastic properties of dough and to initiate yeast growth [38,39]. It is also used as an intermediate in the production of hydrogen peroxide. For example, the oxidation of propane produces a mixture of hydrogen peroxide, methyl alcohol, and formaldehyde. Separation of hydrogen peroxide from the mixture cannot be achieved via distillation unless $Ca(OH)_2$ is added and calcium peroxide is separated in the process. Hydrogen peroxide is regenerated by treating the suspension with carbon dioxide [40-42]. This method is also used for separating hydrogen peroxide solutions from iron and certain other inorganic substances which can be deposited with calcium peroxide [43]. Other applications are the stabilization of vulcanized copolymers of isobutylene [44] and the vulcanization of butyl rubber. Strontium peroxide is also used in the latter application, but other peroxides such as BaO_2, MgO_2, and $K_2S_2O_8$ are not suitable for this purpose [45,46]. A very active calcium oxide can be obtained by heating a mixture consisting of 31.3 wt.% CaO_2 and 68.7 wt.% $Ca(OH)_2$ at 450°C [46a]. In 1959, the price in the United States of one kilogram of 75% calcium peroxide was approximately 7 rubles [47].

Strontium Peroxide – SrO_2

Strontium peroxide, like CaO_2, is obtained by dehydration of the octahydrate. It can be synthesized by the reaction of strontium oxide hydrate solution with a dilute solution of hydrogen peroxide [48,49], or via strontium salt solutions, ammonia, and hydrogen peroxide [50]. The industrial product, synthesized by the first method in the United States, contains 95% SrO_2 [10]. A similar British product contains 82% SrO_2 [11]. A laboratory method is known which is based on the reaction $2SrO + O_2 \rightarrow 2SrO_2$ and is carried out at 500°C and an oxygen pressure of 150 atm [51]. Another method of producing

TABLE VI

$Sr(OH)_2-H_2O_2-H_2O$ System

Solid phase	H_2O_2 content in liquid phase, wt. %				
	$+50°C$	$+30°C$	$+20°C$	$0°C$	$-10°C$
$Sr(OH)_2 \cdot 8H_2O$	0	0−0.38	0−0.72	0−0.34	−
$SrO_2 \cdot 8H_2O$	Trace−2.02	0.38−4.67	0.72−6.2	0.34−10.2	−
$SrO_2 \cdot H_2O_2$	2.89−12.49	−	−	−	−
$SrO_2 \cdot 2H_2O_2$	13.0−90.3	4.67−81.4	6.2−90.7	10.2−86.2	19.12−86.71

strontium peroxide is based on the reaction of strontium nitrate with KO_2 or NaO_2 in liquid ammonia at −30°C, but it has no practical application [19].

In selecting the optimal conditions for the synthesis of strontium peroxide, using the methods in references [48−50], the data on solubility in the $Sr(OH)_2-H_2O_2-H_2O$ system for the temperatures −10, 0, +20, +30, and +50°C (see Table VI) [52], along with the data obtained from thermal studies of $SrO_2 \cdot 8H_2O$ [3, 53], are quite important.

The heating curve of $SrO_2 \cdot 8H_2O$ (see Fig. 13) indicates the presence of three endotherms at 70, 95, and 480°C. The first two correspond to gradual dehydration; the third indicates the decomposition of SrO_2 to $SrO + \frac{1}{2}O_2$ [3, 53]. $SrO_2 \cdot 8H_2O$ crystallizes in the tetragonal system. The lattice constants are: a = 6.340 ± 0.002 A, c = 11.188 ± 0.004 A; Z = 2. Its density is 1.95 g/cm^3 and the x-ray density is 1.947 g/cm^3 [50]. The $SrO_2 \cdot 8D_2O$ compound has been synthesized by mixing a solution containing 0.2 g of strontium nitrate in 2 ml D_2O (99%) with 0.1 g of sodium peroxide in 2 ml of D_2O solution [54].

$SrO_2 \cdot H_2O_2$ and $SrO_2 \cdot 2H_2O_2$ compounds are not very stable. The first compound loses the oxygen of the bound hydrogen peroxide at 65°C, it is dehydrated at 105−110°C, and decomposes to $SrO + \frac{1}{2}O_2$ at 475°C [53]. The second compound liberates the oxygen of the bound hydrogen peroxide at 83°C, is dehydrated at 92°C, and decomposes to $SrO + \frac{1}{2}O_2$ at 404−

491°C [55, 53].   If $SrO_2 \cdot 2H_2O_2$ is decomposed at 50°C (i.e., below the disintegration temperature of the bound hydrogen peroxide) and at a residual pressure, 10 mm Hg, a mixture of SrO_2, $Sr(O_2)_2$, and $Sr(OH)_2$ is obtained [55, 36].

X-ray structural data of strontium peroxide diperoxyhydrate monocrystals are of significant interest.   It has been established [56, 57] that $SrO_2 \cdot 2H_2O_2$ has two modifications: α and β belonging to the monoclinic system. Lattice parameters of the α-modification are: a = 8.262 ± 0.006 A, b = 6.024 ± 0.004 A, c = 8.050 ± 0.005 A.   Its x-ray density is 3.164 g/cm^3, and the optical axis angle is 100°32' ± 8'. Parameters of the β-modification are: a = 7.715 ± 0.004 A, b = 8.754 ± 0.005 A, c = 6.015 ± 0.003 A.   Its x-ray density is 3.074 g/cm^3 and the optical axis angle is 93°40' ± 2'.

When heated, SrO_2 dissociates to Sr + $\frac{1}{2}O_2$ in the temperature range of 410 to 450°C [3, 53]. For the reaction $\frac{1}{2}$Sr + $\frac{1}{2}O_2$ → $\frac{1}{2}SrO_2$: $\Delta F°_{298}$ = −70.6 kcal, $\Delta H°_{298}$ = −76.6 ± 2 kcal, $\Delta S°_{298}$ = −20 eu [28].   The value $\Delta H°_{298}$ = −75.4 ± 0.5 kcal given in reference [58] is probably low, as is the value $S°_{298}$ = 13 eu suggested in reference [29]. SrO_2 is diamagnetic, χ_m = −32.8 · 10^{-6} [30], and its density is 4.7 g/cm^3 [41]. The unit cell is tetragonal with constants a = 3.57 A, c = 6.63 A, c/a = 1.86; Z = 2.   The x-ray density is 4.71 g/cm^3, and the space group is D_{4h}^{17} [41]. The lattice energy is estimated to be 872.5 kcal/mole [59], but according to reference [59a], the value is 635.1 kcal/mole.

Like barium peroxide, strontium peroxide is used in inflammable compositions for tracer bullets and artillery pieces [60, 61].   Strontium peroxide can also be used as a starting substance for obtaining elemental strontium. The peroxide is decomposed to the oxide in a nitrogen atmosphere and is then reduced with aluminum to give metallic strontium [62].

Barium Peroxide − BaO₂

Production Methods. Barium peroxide, the first known peroxide compound, was described by A. Humboldt in 1799 [9]. Today, barium peroxide is no longer used as a starting compound for obtaining hydrogen peroxide by reaction with diluted aqueous acid solutions (sulfuric or phosphoric). With respect to its active oxygen content, barium peroxide is inferior (9.4 wt.%) to the other peroxides of the first and the second groups. However, its relatively inexpensive method of production allows it to be used in many applied fields.

Barium peroxide is obtained by oxidizing BaO with air in a muffle furnace at 500-520°C in the presence of a small amount of water vapor (partial pressure 4-7 mm Hg [63-65]). If the barium oxide is dry, thoroughly crushed, and highly porous and the air is free from CO_2 and excess moisture, the obtained product can contain as much as 88-90% BaO_2 [11-63]. The main criterion for obtaining a high BaO_2 content is the use of sufficiently porous barium oxide. For this purpose, BaO is obtained by heating barium carbonate. Viterite or deposited $BaCO_3$ is heated with a mixture of carbon powder in a pseudo-liquified BaO layer at 700-1000°C. The excess barium oxide prevents sintering of the particles and the formation of an eutectic alloy of BaO and $BaCO_3$. The pseudo-liquified layer is produced by air containing CO_2 which is liberated in the decomposition of barium carbonate, and CO_2 formed by the reduction of the $BaCO_3$ by carbon. Moreover, the formation of CO reduces the partial pressure of CO_2 and enables the heating process to be conducted at a temperature lower than the $BaCO_3$ dissociation temperature (1450°C) [63-65].

Using the heavy oxygen isotope O^{18}, the mechanism of the barium oxide oxidation reaction to barium peroxide has been investigated [66]. The catalytic effect of water on the reaction is explained by the formation of $Ba(OH)_2$ which then undergoes a dehydration according to the following:

$$Ba^{2+} \begin{matrix} \overline{OH} \\ \\ \overline{OH} \end{matrix} + 2O_2 \;\rightarrow\; Ba^{2+} \begin{matrix} \overline{O}\overset{+}{H}\overline{O}_2 \\ \\ \overline{O}\overset{+}{H}\overline{O}_2 \end{matrix} \;\rightarrow\; Ba^{2+} \begin{pmatrix} O \\ | \\ O \end{pmatrix}^{2-} + H_2O + 1.5O_2$$

The unstable, intermediate complex is formed as a result of an electron transfer reaction by which O_2^- ions are formed and as a result of the breaking of the $\overline{O}$:H covalent bonds. Free valences are formed in the oxygen atoms of the hydroxyl groups, and mutual saturation of the valences results in the formation of the peroxide ion with the concurrent decomposition of the intermediate complex.

Barium peroxide can also be formed by oxidizing barium oxide hydrate [67]. The oxidation of barium amalgam [18] has no practical value and results in the formation of a product containing only 60 wt.% BaO_2.

Dehydration of $BaO_2 \cdot 8H_2O$ or $BaO_2 \cdot 2H_2O_2$, synthesized by the reaction of barium oxide hydrate with hydrogen peroxide or from solutions of barium chloride or barium nitrate, ammonium oxide hydrate, and hydrogen peroxide, also produces barium peroxide. In this method, the mechanism for the formation of $BaO_2 \cdot 8H_2O$ differs from that of the BaO-oxygen reaction. Tagged oxygen studies revealed that the mechanism of the former reaction is similar to the formation of an insoluble salt:

$$Ba^{2+} \cdot aq + O_2^{2-} \cdot aq \;\rightarrow\; BaO_2 \cdot 8H_2O \;\; [66]$$

In selecting the optimal conditions for the synthesis of obtaining $BaO_2 \cdot 8H_2O$ and $BaO_2 \cdot 2H_2O_2$, and for the subsequent dehydration of these compounds to BaO_2, it is important to have data concerning the formation of these compounds in the $Ba(OH)_2$–H_2O_2–H_2O system at -10, 0, $+20$, and $+50°C$ [68], as well as thermal stability data [3, 69].

Table VII shows the limiting values of the equilibrium concentrations of hydrogen peroxide for the solid phase regions in the $Ba(OH)_2$–H_2O_2–H_2O system at the above-mentioned temperatures. The BaO concentration in the liquid phases of this system varies from 0.09 to 1.7 wt.%.

According to Table VII, $BaO_2 \cdot 8H_2O$ exists within narrow temperature-concentration limits. It is obvious that it is not economical to produce barium peroxide from the octahydrate under industrial conditions. On the other hand, barium peroxide diperoxyhydrate exists over a wide range of temperatures and hydrogen peroxide concentrations. For this reason, a method was introduced to obtain barium peroxide [13] based on the dehydration of $BaO_2 \cdot 2H_2O_2$ at 100-115°C and at atmospheric pressure. Diperoxyhydrate is obtained at room temperature by adding a barium nitrate solution to a mixture of water, hydrogen peroxide, and ammonium oxide hydrate containing 3.5 to 4% H_2O_2, and a $Ba(NO_3)_2:H_2O_2$ ratio of 1:2.5. It has also been suggested that barium peroxide be obtained by the dehydration of the monoperoxyhydrate at 200°C. However, a patent claim [70] that the initial $BaO_2 \cdot H_2O_2$ can be obtained at 5-15°C does not agree with the data presented in Table VII. According to this table and Figs. 10 and 11 [68], barium peroxide monoperoxyhydrate in the $Ba(OH)_2-H_2O_2-H_2O$ system exists only at 50°C. By applying the above-mentioned information, the author of this book, together with employees from one plant, has introduced a method for obtaining BaO_2 from solutions of $Ba(OH)_2$ and H_2O_2 instead of the time consuming method based on the oxidation of BaO.

Barium peroxide is very inert to water, and it can be readily separated from soluble compounds [71], and even from barium oxide [72].

It should be noted that in carrying out investigations of metal hydroxide–hydrogen peroxide–water systems, such as the $Ba(OH)_2$ system, or the previously described LiOH, $Ca(OH)_2$, and $Sr(OH)_2$ systems, it is necessary for the investigators to consider the "optimal reaction," or the "equi-

TABLE VII

$Ba(OH)_2-H_2O_2-H_2O$ System

Solid phase	H_2O_2 content in liquid phase, wt. %			
	−10°C	0°C	+20°C	+50°C
$BaO_2 \cdot 8H_2O$	—	0.23−2.0	0−0.9	—
$BaO_2 \cdot H_2O_2$	—	—	—	0−7.9
$BaO_2 \cdot 2H_2O_2$	9.8−89.6	2.0−90.6	0.9−84.8	7.9−64.0

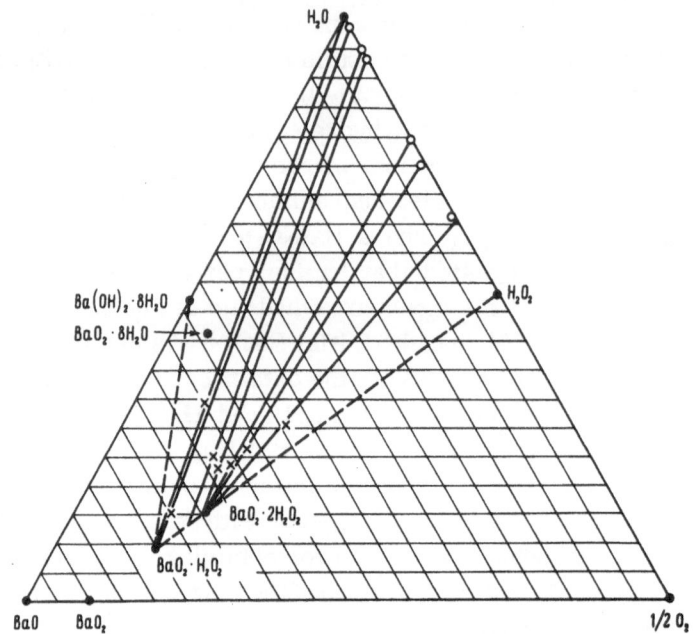

Fig. 10. Solubility isotherm at $+50°C$ for the $Ba(OH)_2$–H_2O_2–H_2O system (in triangular coordinates) [68].

librium time," i.e., the time span during which no noticeable decomposition of the peroxide solid phase and the hydrogen peroxide solution takes place. Figure 12 shows the data obtained by N. K. Grigor'eva [68] from a study of the $Ba(OH)_2$–H_2O_2–H_2O system using initial hydrogen peroxide solutions of 3.74 and 8.45 wt.% concentration. Based on a constant composition of hydrogen peroxide and barium oxide in the liquid phase, the "equilibrium time" was 2.5-3.5 hr. Keeping the solution in contact with the solid phase for a longer time is not desirable because of the possibility of hydrogen peroxide decomposition which would distort the equilibrium of the system.

Properties of Barium Peroxide. As seen in Table VII, barium peroxide forms a series of molecular compounds with water and hydrogen peroxide. $BaO_2 \cdot 8H_2O$ is quite stable. Its heating

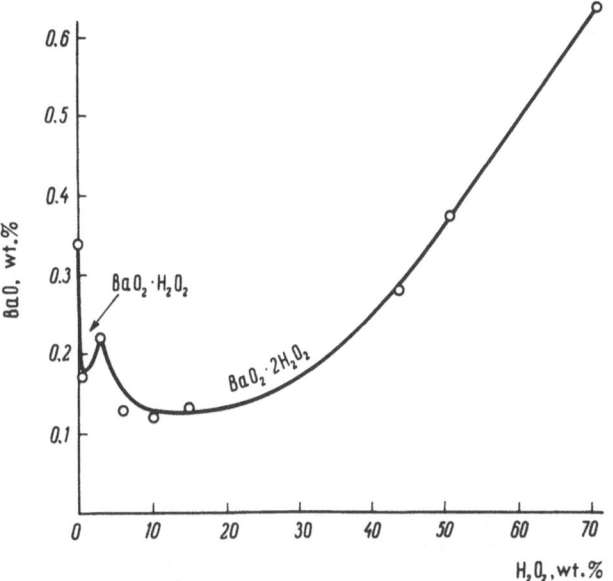

Fig. 11. Solubility isotherm at +50°C for the $Ba(OH)_2 - H_2O_2 - H_2O$ system.

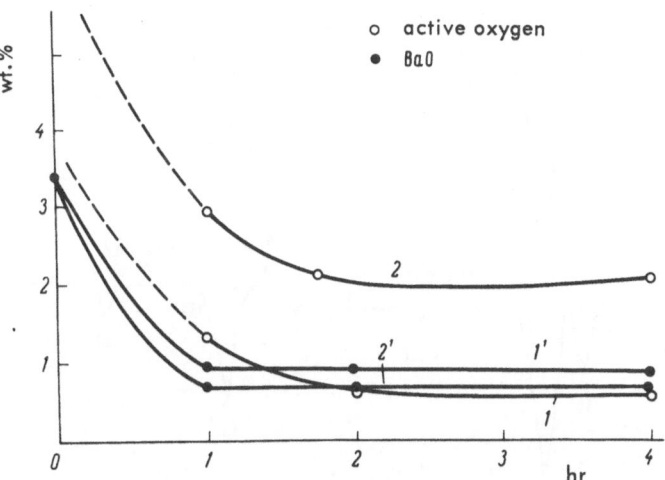

Fig. 12. Equilibrium at +20°C in the $Ba(OH)_2 - H_2O_2 - H_2O$ system for initial H_2O_2 concentrations: 1,1') 3.74 wt.%; 2,2') 8.45 wt.%.

curve (see Fig. 13) shows three endotherms at 40-55, 100, and 790°C. The first corresponds to partial melting, the second to dehydration, and the third to BaO_2 decomposition to BaO and $\frac{1}{2}O_2$ [3, 69]. The unit cell of $BaO_2 \cdot 8H_2O$ is tetragonal with parameters a = 6.51 A, c = 11.50 A [72a]. The O—O distance in $BaO_2 \cdot 8H_2O$ is 1.48 A. The identical bond length is found in $SrO_2 \cdot 8H_2O$, $CaO_2 \cdot 8H_2O$, anhydrous BaO_2, and H_2O_2 [73].

$BaO_2 \cdot H_2O_2$ and $BaO_2 \cdot 2H_2O_2$ compounds are not very stable. Starting at 60-75°C they lose the oxygen from the bound hydro-

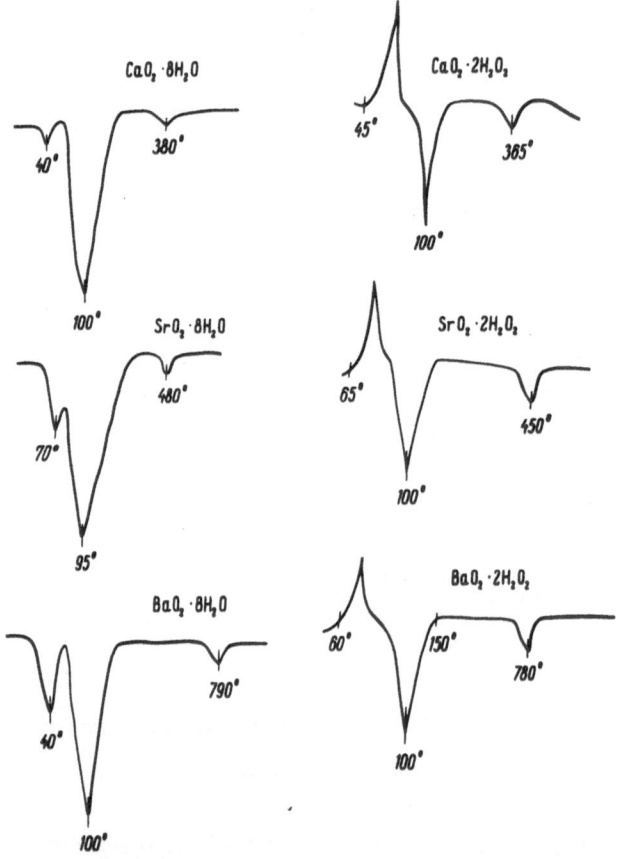

Fig. 13. Differential heating curves for octahydrates and diperoxyhydrates of calcium, strontium, and barium peroxides.

TABLE VIII

Crystallochemical Constants of $BaO_2 \cdot 2H_2O_2$

	α-$BaO_2 \cdot 2H_2O_2$	β-$BaO_2 \cdot 2H_2O_2$	γ-$BaO_2 \cdot 2H_2O_2$
a, A	8.454 ± 0.006	8.18 ± 0.03	16.837 ± 0.014
b, A	6.398 ± 0.004	9.03 ± 0.04	6.407 ± 0.006
c, A	8.084 ± 0.005	6.35 ± 0.03	8.005 ± 0.007
β	96°,32 ± 4'	95°,40 ± 18'	–
V, cm³	434.4	467	863.5
Density, g / cm³	3.57	3.41	3.56
X-ray density, g/ cm³	3.629	3.38	3.651
Z	4	4	8

gen peroxide, and dehydration starts at 100°C [69]. $BaO_2 \cdot H_2O_2$ crystals belong to the monoclinic system with parameters a = 4.132 ± 0.004 A, b = 9.464 ± 0.009 A, c = 8.308 ± 0.008 A; β = 98°35 ± 5'. The density is 4.04 g/cm³, and the x-ray density is 4.19 g/cm³ [74]. The $BaO_2 \cdot H_2O_2 \cdot 2H_2O$ compound has also been synthesized and also belongs to the monoclinic system with parameters a = 10.39 ± 0.04 A, b = 8.36 ± 0.03 A, c = 6.58 ± 0.03 A; β = 95°40 ± 18' [75]. $BaO_2 \cdot 2H_2O_2$ exists in three modifications: α, β, and γ [75], the lattice parameters of which are shown in Table VIII. The decomposition of $BaO_2 \cdot 2H_2O_2$ at 50°C and at a residual pressure of 10 mm Hg yields a product containing $Ba(O_2)_2$, together with BaO_2 and $Ba(OH)_2$ [76]. The peroxide oxygen of $BaO_2^{18} \cdot 8H_2O$ can be exchanged with hydrogen peroxide solutions of natural isotopic composition and with concentrations exceeding 1 wt.% to form $BaO_2 \cdot 2H_2O_2$, in which all the peroxide oxygen atoms have the same isotopic composition [37].

Barium peroxide is the most thermally stable of all the known metal peroxides. If heated at atmospheric pressure, it extensively liberates oxygen only at temperatures above 500°C. The exotherm in its heating curve [3] at 790°C corresponds to the decomposition of barium peroxide to barium oxide. Because of the continuous increase in the decomposition pressure at higher temperatures, its melting point has not been determined. However, the formed substance melts at approximately 800°C and loses all of its active oxygen at 900°C [63, 69]. Dissociation kinetics of BaO_2 have been determined in the temperature range of 500 to 700°C and at pressures from $1 \cdot 10^{-3}$ to 203 mm Hg [21, 22]. The apparent energy of activation for the dissociation process has been estimated as 45,800 kcal/mole. The variation of the decomposition pressures indicates the formation of solid solutions. The decomposition of BaO_2 in a high vacuum takes place in the temperature range from 410 to 460°C [23,24]. Barium peroxide octahydrate decomposes at a higher rate in high vacuum and at a significantly lower temperature than anhydrous BaO_2 [23, 24].

The density of BaO_2 is 5.43 g/cm^3 [34, 27]. The bulk density of BaO_2 obtained by oxidation of BaO is 2.1 g/cm^3, whereas the bulk density of BaO_2 obtained from $Ba(OH)_2$ and H_2O_2 is 0.79 g/cm^3. Its basic cell is tetragonal with parameters a= 3.807 A, c = 6.83 A, c/a = 1.80; Z = 2. X-ray density is 5.43 g/cm^3. Molecular volume is 29.9 cm^3. The space group is D_{4h}^{17} [27]. In the BaO_2 cell, the distances between the barium and oxygen atoms are not the same for the different planes; they are 2.68 and 2.79 A, respectively [77]. BaO_2 is diamagnetic, $\chi_m =$ $-40.6 \cdot 10^{-6}$ [78]. The BaO_2 lattice energy is estimated as 823 kcal/mole [59].

For the reaction $Ba + O_2 \rightarrow BaO_2$, $\Delta H°_{298} = -151.89 \pm 0.25$ kcal [58], $\Delta F°_{298} = -140$ kcal, $\Delta S°_{298} = -20$ eu [28]. For the reaction $BaO + \frac{1}{2}O_2 \rightarrow BaO_2$, $\Delta H°_{298} = -17.74 \pm 0.25$ kcal [79]. The average molecular heat capacity for BaO_2 in the temperature range from 19 to 100°C is approximately 16 kcal/mole-deg [80]. The reported value [29] for $S°_{298} = 15.7$ eu is incorrect. The values of 22.62 [81] and 24.6 eu are more likely [21], but the most likely value is 25.1 eu [82].

Barium peroxide reacts with oxygen at 210°C and 3300 atm forming $Ba(O_2)_2$ [82]. In reducing barium peroxide with hydrogen, at temperatures up to 300°C, $Ba(OH)_2$ is formed. At higher temperatures, barium peroxide is reduced to barium oxide [21, 22]. The reduction rate of barium peroxide with hydrogen initially follows the monomolecular reaction equation [23].

Studies of the reaction of BaO_2 with carbon dioxide have revealed that the reaction takes place only in the presence of water vapor. Barium peroxide octahydrate reacts more readily with carbon dioxide than the anhydrous peroxide [22]. With NO_2, barium peroxide reacts at temperatures of 0 to 400°C with the formation of barium nitrate [83]. With sulfur, barium peroxide reacts according to the following equations [84, 85]:

$$S + BaO_2 \rightarrow BaS + O_2$$

$$2S + 2BaO_2 \rightarrow BaSO_4 + BaS$$

$$S + 3BaO_2 \rightarrow BaSO_4 + 2BaO$$

The ignition temperature of these mixtures is 250°C. With selenium, barium peroxide reacts according to the following reactions:

$$Se + 3BaO_2 \rightarrow BaSeO_4 + 2BaO + 95.3 \text{ kcal}$$

$$2Se + 2BaO_2 \rightarrow BaSeO_4 + BaSe + 121 \text{ kcal}$$

Mixtures of selenium and barium peroxide ignite at approximately 265°C. Mixtures containing up to 15 wt.% Se generate heat to approximately 500-600°C, and mixtures containing over 15 wt.% Se generate heat to 600-1000°C [85a]. Barium peroxide reacts with molybdenum to form barium molybdate [85] according to the equation

$$Mo + 3BaO_2 \rightarrow BaMoO_4 + 2BaO$$

The reaction with iron is as follows:

$$2Fe + 3BaO_2 \rightarrow Fe_2O_3 + 3BaO \text{ [85, 86]}$$

It has been established that Fe_2O_3 accelerates the liberation of oxygen from barium peroxide. In the process, Fe_2O_3 acts as a catalyst up to 500°C, and over 600°C it reacts with BaO_2 forming $BaO \cdot 6Fe_2O_3$ [87]. A mixture of metallic magnesium and barium peroxide ignites at 548°C. A mixture of calcium resinate and barium peroxide ignites at 278°C, and a ternary mixture of Mg–BaO_2–calcium resinate ignites at 299°C [88]. BaO_2 does not react with CuO, but it reacts with Cu_2O at 260°C forming CuO and BaO. It reacts with manganese dioxide to form $BaMnO_4$ and BaO, and with Cr_2O_3 to form $BaCrO_4$ [89]. A mixture of barium peroxide and any manganese oxide is transformed to $Ba_3(MnO_4)_2$ when heated in a stream of dry oxygen; in a stream of moist oxygen, it is transformed to $Ba_5(MnO_4)_3OH$ [90].

SiO_2 and BaO_2 mixtures react, no matter what the initial components ratio, to form metasilicate or the salts of the stronger silicate acids $BaSi_2O_5$ and $BaSi_3O_5$. The orthosilicate radical is formed only on heating to higher temperatures. In mixtures enriched with barium peroxide, tribarium silicate begins to form at 750-800°C, i.e., after decomposition of the peroxide. In mixtures enriched in SiO_2, the basic silicates which form react further at high temperatures with the excess SiO_2 to form more acidic silicates [91]. Equations have been derived for the change of free energy as a function of temperature for the following reactions:

$$BaO_2 + SiO_2 \rightarrow BaSiO_3 + \tfrac{1}{2}O_2$$

$$2BaO_2 + SiO_2 \rightarrow Ba_2SiO_4 + O_2$$

$$3BaO_2 + SiO_2 \rightarrow Ba_3SiO_5 + \tfrac{3}{2}O_2$$

$$BaO_2 + 2SiO_2 \rightarrow BaSi_2O_5 + \tfrac{1}{2}O_2$$

$$BaO_2 + BaSiO_3 \rightarrow Ba_2SiO_4 + \tfrac{1}{2}O_2 \text{ [92]}$$

Thermal effects have been measured for the reaction

$$3BaO_2 + 2Al \rightarrow Al_2O_3 + 3BaO + 352 \text{ kcal [32]}$$

and the feasibility of utilizing barium peroxide as a source of heat in metallurgical processes has been demonstrated.

With anhydrous HF, barium peroxide reacts by liberating oxygen [93] via the following reaction:

$$BaO_2 + 2HF \rightarrow BaF_2 + H_2O + \tfrac{1}{2}O_2$$

whereas with anhydrous HCl, the following reaction occurs:

$$3BaO_2 + 8HCl \rightarrow 3BaCl_2 + 4H_2O + Cl_2 + O_2$$

With aqueous solutions of the same acids, barium peroxide forms hydrogen peroxide:

$$BaO_2 + 2HX \rightarrow BaX_2 + H_2O_2$$

where X = halogen.

Applications of Barium Peroxide. Barium peroxide is used as the starting material for obtaining hydrogen peroxide by means of its reaction with diluted acid solutions [63]. It is mixed with catalysts as a promoter for the following purposes: the polymerization of olefins [94] and vinyl chloride [95]; the oxidation of ethylene to ethylene oxide [96-99]; the production of synthetic gas ($CO + H_2$) by partial oxidation of hydrocarbons (methane or natural gas) [100]; cracking high boiling-point hydrocarbons [101]; and the synthesis of hydrazine hydrate from ammonia and oxygen [102]. Barium peroxide is also used in combination with polytrifluorochloroethylene to improve its thermal properties [103]. It is also used in pyrotechnology [10] as a component of a mixture in incendiary bullets [104], tracer bullets, artillery pieces [60, 61, 105-107], in delayed igniting composites [108-112], detonating capsules [85a, 88], and in water resistant fuses [113]. In delayed igniting composites, barium peroxide is used along with powdered manganese [88], selenium or tellurium [109], iron potassium permanganate [114], and other substances. It is also used in the decoloring of lead [115] and tellurium glasses [116], as a component in mixtures used in coating thermionic cathodes [117-124], to obtain porous plastics [125], and as a starting substance for the production of metallic barium [62, 126]. In the last application, the mixture, consisting of 100 parts of BaO_2, 100 parts of CuO or metallic copper, and 129 parts of powdered aluminum, is pressed and heated in a vacuum at 300–600°C. After 3 hr the temperature is raised to 1350°C. The sublimated barium vapors are deposited on the cold walls of the reaction vessel [126]. Barium compounds can be separated from iron by the use of barium peroxide [126]. Barium peroxide is also used in a mixture with triethylene oxide for producing disinfecting vapor substances [128]. In analytical chemistry, barium peroxide can be added to sulfuric acid in order to accelerate the combustion of organic substances [129].

Barium peroxide is harmful to the respiratory organs. The tolerable concentration is 0.5 mg per cubic meter of air [130].

In 1959, the price of one kilogram of BaO_2 in the United States was 1.85 rubles [47]. Requirements for the domestic use of barium peroxide are presented in reference [131].

TABLE IX

Basic Properties of Alkaline Earth Metal Peroxides

Parameter	Formula		
	CaO_2	SrO_2	BaO_2
Active oxygen content, wt.%	22.2	13.4	9.4
Density, g/cm^3	2.92	4.7	5.43
Complete disintegration temp., °C	375–425	410–450	790–900
ΔH°_{298}, kcal/mole	-156.4 ± 1	-153.2 ± 4	-151.89 ± 0.25
ΔF°_{298}, kcal/mole	-144	-141.2	-140
ΔS°_{298}, eu	(-20)	(-20)	(-20)
ΔH°_{298} (from oxide and oxygen), kcal/mole	-5.8 ± 0.5	-12.1 ± 2	-17.1
$\chi_m \cdot 10^6$	-23.8	-32.8	-40.6
Lattice	tetragonal	tetragonal	tetragonal
U,* kcal/mole	(706.5)	(664.6)	(635.1)

*Reference [59a].

Table IX shows the basic properties of alkaline earth metal peroxides. It can be seen from the data presented in the table that the heats of formation of the alkaline earth metal peroxides follow the thermodynamic logarithmic rule in the same manner as the metal peroxides. It is possible [5], therefore, to determine by extrapolation the heat of formation of radium peroxide, which has not yet been prepared in pure form.

Figure 13 shows heating curves for the octahydrates and diperoxyhydrates of calcium, strontium, and barium peroxide. It is seen that the exotherm associated with the decomposition of the bound hydrogen peroxide occurs before the endotherm associated with dehydration which is typical for this type of compound.

MAGNESIUM PEROXIDE AND PEROXIDES OF THE ZINC SUBGROUP METALS

Magnesium Peroxide – MgO_2

The question concerning the existence of various magnesium peroxide compounds is discussed in detail in reference [132]. By studying the solubility in the $Mg(OH)_2-H_2O_2-H_2O$ system at 0 and 20°C, the feasibility of the existence of MgO_2, $MgO_2 \cdot 0.5H_2O$, and $MgO_2 \cdot H_2O$ in equilibrium with H_2O_2 has been demonstrated. Table X shows the equilibrium concentration limits of hydrogen peroxide for the solid phase regions in the $Mg(OH)_2-H_2O_2-H_2O$ system at the above-mentioned temperatures.

To date, however, 100% MgO_2 has not been produced. Treatment of $Mg(OH)_2$, under laboratory conditions, with a large excess of 80% H_2O_2 solution, followed by drying of the residue over P_2O_5 and KOH at room temperature, produces products containing 80% by weight of MgO_2, i.e., with the approximate composition $MgO_2 \cdot 0.5H_2O$. In some studies [133, 54], the existence of MgO_2 hydrates has been discounted. Anhydrous magnesium peroxide is considered to have a cubic lattice with the following parameters: a = 4.839 ± 0.007 A,

$Z = 4$ [133]. The heating curve of hydrated MgO_2 (see Fig. 14) displays an endotherm at 100°C which corresponds to dehydration, and an exotherm at 375°C which corresponds to decomposition to MgO [2, 3]. For the $\frac{1}{2}Mg + \frac{1}{2}O_2 \rightarrow \frac{1}{2}MgO_2$ reaction, $\Delta F°_{298} = -68$ kcal, $\Delta H°_{298} = -74.4 \pm 0.5$ kcal, and $\Delta S°_{298} = -20$ eu [28].

The industrial method of producing MgO_2 is based on the reaction of magnesium oxide with hydrogen peroxide and yields products containing not more than 50% MgO_2. An electrochemical method which was suggested by F. Hinz [134] has found no practical use. The reaction of magnesium salts with ammonia solutions and hydrogen peroxide produces deposits which are quite unstable, difficult to filtrate, and which have low MgO_2 content. Another method for obtaining 25% MgO_2, based on a reaction of magnesium nitrate with KO_2 or NaO_2 in liquid ammonia at −30°C, has no practical application [19]. The author of this book has developed and verified under industrial conditions a method for obtaining a product containing up to 60 wt.% MgO_2. The method is based on the reaction of milk of magnesia with hydrogen peroxide [135]. The process consists of four steps: 1) preparation of milk of magnesia containing 7 wt.% MgO, and boiling of the aqueous MgO suspension; 2) reaction of one volume of milk of magnesia with one volume of hydrogen peroxide at 15°C for one day; 3) filtration in vacuum; 4) drying in a vacuum at 70°C, with a residual pressure of 80 mm Hg for one day.

TABLE X

$Mg(OH)_2 - H_2O_2 - H_2O$ System

Solid phase	H_2O content in liquid phase, wt.%	
	0°C	20°C
$Mg(OH)_2$	0−1.5	0−2.0
$MgO_2 \cdot H_2O$	1.5−62	−
$MgO_2 \cdot 0.5H_2O$	62−97	2−30.0
MgO_2	−	30−62

Magnesium peroxides, classified as "heavy," "semi-heavy," and "light," are produced in a number of countries. They find application mainly in medicine, for counteracting hyperacidity in the gastric intestinal tract, in treating such metabolic diseases as diabetes and ketonuria, and in the preparation of antiseptic ointments, toothpaste, and powders. According to literature sources, magnesia products based on magnesium peroxide can be used for sterilization of water in the field and for bleaching cotton. All the above-mentioned products represent a mixture of peroxide, oxide, and magnesium oxide hydrate with an admixture of magnesium carbonate in which the content of MgO_2 does not exceed 50 wt.%. For example, products manufactured in Great Britain [11] contain 25 wt.% and in the United States, 50 wt.% MgO_2 [10]. A description of magnesium peroxide products manufactured in the USSR is given in reference [130].

Zinc Peroxide − ZnO_2

As a result of studies of the $Zn(OH)_2−H_2O_2−H_2O$ system at temperatures from −20° to +30°C [136, 137, 138], the presence of hydrated solid phases of undetermined composition and nature has been established. Only $ZnO_2 \cdot 0.5H_2O$ could be recovered from the system [139]. X-ray structural analysis and the infrared spectrum [54] of the reaction products obtained from the reaction of ZnO with hydrogen peroxide did not confirm the presence of compounds containing any water of hydration. Zinc peroxides have a cubic lattice with a = 4.871 ± 0.006 A [140]. For a cubic lattice with Z = 4, the x-ray density is 5.6 g/cm^3 [140].

The heat of formation of ZnO_2 is estimated to be 83 ± 15 kcal/mole [140a]. The heating curve of hydrated zinc peroxide (see Fig. 14) indicates the presence of an endotherm at 115°C corresponding to dehydration, and an exotherm at 205-225°C corresponding to decomposition to ZnO [13, 139]. The information given in reference [140b] indicating that the decomposition of ZnO_2 at 200°C is an endothermic process is incorrect.

In references [136,139], methods are described for producing a material containing up to 76 wt.% ZnO_2 either from a suspension of zinc oxide hydrate and a solution of hydrogen peroxide [140], or from a solution of zinc salt, ammonia, and hydrogen peroxide [4]. It has also been established that products containing up to 70 wt.% ZnO_2 can be obtained from the reaction of zinc nitrate with KO_2 or NaO_2 in liquid ammonia at −30°C [19]. Industrial methods for obtaining zinc peroxide products are based on the reaction of zinc oxide at 30-70°C with a 15% hydrogen peroxide solution containing 2-2.5 ml/liter of mineral acids [142], or via the reaction of a suspension of zinc oxide hydrate with a 5% hydrogen peroxide solution at room temperature [141]. A stable zinc peroxide compound is obtained by drying the hydrated product, obtained by any of the above-mentioned methods, at 100°C at atmospheric pressure for 4 to 5 hr [143], or by vacuum drying at 70-80°C for 3 hr [141].

The industrial product manufactured in Great Britain and in the United States [10] contains 55 wt.% ZnO_2. Its price in 1959 was approximately 7 rubles per kilogram [47]. Zinc peroxide is used mainly in the cosmetic industry and for medical purposes [10,144], but it also finds use as an inhibitor in the accelerated vulcanization of rubber [145] and as a filler in siloxane elastomers [146].

In recent years, a series of zinc peroxyorganic compounds have been synthesized [147-149] by the rapid oxidation of organo-zinc compounds such as diethyl- and di-n-butyl-zinc. Typical reactions are

$$R_2Zn + O_2 \rightarrow RZnOOR$$

$$RZnOOR + O_2 \rightarrow Zn(OOR)_2$$

The products are solid diamagnetic substances, which are rapidly hydrolyzed in moist air and detonate when heated.

Cadmium Peroxide — CdO_2

X-ray investigations of the solid phases formed in the reaction of cadmium nitrate solutions with hydrogen peroxide and sodium oxide hydrate indicate the formation of only one solid phase at 20 and 80°C. The phase is of variable composition with maximum limits $Cd(O_2)_{0.74}(OH)_{0.52}(H_2O)_{2.46}$ to $Cd(O_2)_{0.88}(OH)_{0.24}$. The latter has a cubic structure and is isomorphous pyrite. Its cell parameter is a = 5.273 ± 0.003 A, and its density is 5.45 g/cm^3 [150]. However, according to reference [151], if solutions of the nitrate, acetate, or sulfate of Cd are reacted with 30% hydrogen peroxide and ammonium oxide hydrate at 80°C, followed by drying at 120°C, CdO_2 is formed, which has a cubic structure with a = 5.313 ± 0.003 A, and is isomorphous with pyrite. The density, however, is 5.93 g/cm^3. The heat of formation of CdO_2 is estimated as 60 ± 15 kcal/mole [140a]. In the reaction of $Cd(OH)_2$ with hydrogen peroxide solutions, products have been obtained with a composition $CdO_2 \cdot xH_2O$ (where x = 2 to 0.5) [152]. In the reaction of alkali with solutions of the cadmium salt in hydrogen peroxide, the formed products have the composition $2CdO_2 \cdot CdO$ [153]. The heating curve of $CdO_2 \cdot 2H_2O$ (see Fig. 14) indicates the presence of an endotherm associated with dehydration at 110°C, and an exotherm indicating decomposition at 195°C [2, 3]. The decomposition of CdO_2 at ~ 200°C has been confirmed [151].

Cadmium diperoxyorganic compounds, in the form of white powders stable to 50°C, are obtained from the reaction of CdR_2 compounds such as diethyl cadmium or dibutyl cadmium with an alkyl hydroperoxide, or by the oxidation of CdR_2 compounds, according to the following equations:

$$CdR_2 + 2R'OOH \rightarrow 2RH + Cd(OOR')_2$$

$$CdR_2 + 2O_2 \rightarrow Cd(OOR)_2$$

Under similar conditions, dimethyl cadmium forms $H_3CCdOOH$ [154]. The reaction of equimolar quantities of diaryl cadmium and a 98% solution of hydrogen peroxide at −20°C yields a rather unstable product containing 74.2 wt.% Cd and 16% active oxygen, which corresponds approximately to the formula CdO_2. If the initial substances were reacted in a 2:1 molar ratio, the product obtained was $H_3CCdOOH$ [155, 156].

A comparison of the heating curves of the peroxide octahydrates of calcium, strontium, and barium with the heating curves of hydrated forms of magnesium, zinc, and cadmium peroxides, conforming to the general formula $MO_2 \cdot xH_2O$ (Fig. 14), reveals significant differences. The heating curves of the calcium, strontium, and barium compounds show all

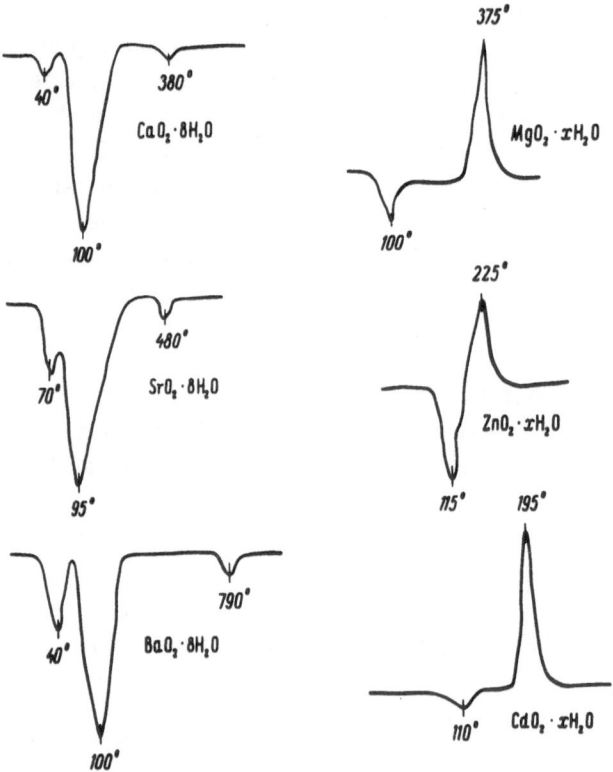

Fig. 14. Differential heating curves of hydrate forms of the group two metal peroxides.

three endotherms, corresponding to the partial melting, dehydration, and decomposition of the peroxides to the oxide and oxygen. In this series, it is found that as the atomic weight of the metal increases, the stability of the peroxide formed also increases. For example, the thermal decomposition of CaO_2 begins at 380°C, SrO_2 at 480°C, and BaO_2 at 790°C.

The heating curves of the hydrated peroxides of magnesium, zinc, and cadmium indicate the presence of only one endotherm at 100-115°C corresponding to dehydration, and one exotherm corresponding to the decomposition of peroxide compound to oxide. In this series, the decomposition temperature decreases as the atomic weight of the metal forming the peroxide compound increases. Thus, for MgO_2 the decomposition temperature is 375°C, for ZnO_2 it is 225°C, and for CdO_2 it is 195°C. As a result of these differences, and the fact that the basicity of the oxides of these elements is lower than that of calcium, strontium, and barium, it can be assumed that the hydrated forms of magnesium, zinc, and cadmium peroxides are not hydrates of true peroxide, $M^{2+}O_2^{2-}$, but rather belong to a group which is intermediate between peroxides of the above-mentioned composition and the peroxyacids. A possible structure of these compounds can be expressed by the formula $OH-M-OOH \cdot xH_2O$.

That the magnesium and zinc compounds exist as this type of peroxide is based on the analysis of the products formed in the reaction of the corresponding oxides with hydrogen peroxide solutions in a NaOH medium [157]. The infrared spectra of these compounds do not indicate the presence of bound water [54].

As a result of a comparison of the heating curves of the hydrated peroxide forms of magnesium, zinc, and cadmium (see Fig. 14) with the heating curves of the diperoxyhydrates of calcium, strontium, and barium (see Fig. 13), the hypothesis that the first compounds contain bound hydrogen peroxide can be discounted since the exothermal effect corresponding to their decomposition does not precede the dehydration effect as is typical for the peroxide peroxyhydrates.

Mercury Peroxide – HgO_2

Evidence for the possible existence of mercury peroxide, HgO_2, and its hydrates was presented in the beginning of this century [9]. The mercury compounds, however, were not investigated systematically until 1959. Today, it is agreed that at low temperatures, the reaction of yellow mercury oxide with hydrogen peroxide solutions produces two HgO_2 modifications, α and β, containing 13.3 wt.% oxygen [158]. α-HgO_2 is yellow and quite unstable. It is formed by the reaction of yellow mercury oxide with 20% hydrogen peroxide solution at $-15°C$. β-HgO_2 is reddish-yellow, more stable, and is formed by the reaction of yellow mercury oxide with 30% hydrogen peroxide solution at temperatures below $-15°C$. α-HgO_2 is rhombohedral with parameters a = 6.080 ± 0.010 A, b = 6.010 ± 0.010 A, c = 4.800 ± 0.010 A [158].

REFERENCES

1. I. I. Vol'nov. Izv. Sektora Fiz.-Khim. Analiza 26:211 (1955).
2. I. I. Vol'nov. Dokl. Akad. Nauk SSSR 94:477 (1954).
3. I. I. Vol'nov. Zh. Neorgan. Khim. 3:538 (1958).
4. L. V. Ladeinova. Izv. Akad. Nauk SSSR, Otd. Khim. Nauk (1959), p. 195.
5. A. F. Kapustinskii. Izv. Akad. Nauk. SSSR, Otd. Khim. Nauk (1948), p. 586.
6. I. I. Vol'nov, V. N. Chamova, V. P. Sergeeva, and E. I. Latysheva. Zh. Neorgan. Khim. 1:1937 (1956).
7. H. Kobayashi. Japanese Patent 175368 (1947).
8. J. H. Young. U.S. Patent 2533660 (1950).
9. Hydrogen Peroxide and Peroxide Compounds, edited by M. E. Pozin, Moscow, Goskhimizdat (1951).
10. R. Kirk and D. Othmer. Encyclopedia of Chemical Technology, Vol. 10, New York, The Interscience Encyclopedia (1953), p. 28.
11. V. W. Slater. Chem. Ind. 5:45 (1945).
12. R. S. Shineman and A. J. King. Acta Cryst. 4:67 (1951).
13. S. Z. Makarov and N. K. Grigor'eva. Zh. Prikl. Khim. 32:2184 (1959).
14. S. Z. Makarov and N. K. Grigor'eva. Izv. Akad. Nauk SSSR, Otd. Khim. Nauk (1954), p. 385.
15. S. Z. Makarov and N. K. Grigor'eva. Izv. Akad. Nauk SSSR, Otd. Khim. Nauk (1958), p. 1289.
16. V. N. Chamova and V. P. Sergeeva. Zh. Neorgan. Khim. 2:1938 (1957).
17. S. Z. Makarov and N. K. Grigor'eva. Izv. Akad. Nauk SSSR, Otd. Khim. Nauk (1954), p. 598.
18. H. Hahn. Z. Anorg. Allgem. Chem. 275:35 (1954).
19. D. Schechter and J. Kleinberg. J. Am. Chem. Soc. 76:3297 (1954).
20. Ch. N. Satterfield and T. W. Stein. Ind. Eng. Chem. 46:1734 (1954).
21. B. D. Averbukh and G. I. Chufarov. Zh. Obshch. Khim. 21:629 (1951).
22. B. D. Averbukh. Candidate Dissertation, Sverdlovsk, UFAN (1948).
23. Ya. S. Rubinchik. Candidate Dissertation, Minsk, BGU (1953).

24. M.M. Pavlyuchenko and Ya. S. Rubinchik. Zh. Fiz. Khim. 32:848 (1958).
25. C. Brosset and N.G. Vannerberg. Nature 177(501):236 (1956).
26. I.I. Vol'nov and A.N. Shatunina. Zh. Neorgan. Khim. 2:1474 (1957).
27. H. Föppl. Z. Anorg. Allgem. Chem. 291:46 (1957).
28. L. Brewer. Chem. Rev. 52:6 (1953).
29. V. Latimer. Oxidation State of Elements, Moscow, IL (1954).
30. A.B. Neiding and I.A. Kazarnovskii. Dokl. Akad. Nauk SSSR 78:713 (1951).
31. M.M. Pavlyuchenko and Ya. S. Rubinchik. Zh. Neorgan. Khim. 4:50 (1959).
32. N.N. Murach and V.K. Kulifeev. Izv. VUZOV, Nonferrous Metallurgy, No. 6:64 (1958).
33. N.G. Vannerberg. The Formation and Structure of Peroxy Compounds of Group IIa and IIb Elements, Göteborg, Uppsala, Almqvist (1959).
34. I.I. Vol'nov, V.N. Chamova, and E. I. Latysheva. Zh. Neorgan. Khim. 2:263 (1957).
35. I.I. Vol'nov and V.N. Chamova. Zh. Neorgan. Khim. 3:1098 (1958).
36. I.I. Vol'nov, V.N. Chamova and V.P. Sergeeva. Zh. Neorgan. Khim. 4:253 (1959).
37. V.A. Lunenok-Burmakina and A.P. Potemskaya. Ukr. Khim. Zh. 28:48 (1962).
38. W. Hundt and K. Wieweg. Seifen-Oele-Fette-Wachse 81:419, 444 (1955).
39. J. Barber and B. Kennedy. Cereal Chem. 35:201 (1958).
40. Ch. N. Satterfield. Ind. Eng. Chem. 46:1007 (1954).
41. G. Cook. Canadian Patent 508635 (1954).
42. Air Liquide, French Patent 1256598 (1960).
43. A.K. Dunlop and D.R. Douslin. U.S. Patent 2695217 (1954).
44. A. Lefevre. French Patent 1032708 (1953).
45. R.L. Zapp. U.S. Patent 2666753 (1954).
46. R.L. Zapp. J. Polymer Sci. 9:97 (1952).
46a. D.R. Glasson. J. Appl. Chem. 13:111 (1963).
47. Chem. Eng. News 37(26):56 (1959).
48. N. J. Rentschler. U.S. Patent 2415443 (1947).
49. N.J. Rentschler. Swedish Patent 147004 (1954).
50. N.G. Vannerberg. Arkiv Kemi 14:17 (1959).
51. C.B. Holtermann. Ann. Chim. 14:121 (1940).
52. S.Z. Makarov and T.I. Arnol'd. Izv. Akad. Nauk SSSR, Otd. Khim. Nauk (1958), p. 1407.
53. S.Z. Makarov and T.I. Arnol'd. Izv. Akad. Nauk SSSR, Otd. Khim. Nauk (1959), p. 774.
54. N.G. Vannerberg. Arkiv Kemi 14:107 (1959).
55. I.I. Vol'nov and E.I. Latysheva. Zh. Neorgan. Khim. 2:259 (1957).
56. N.G. Vannerberg. Acta Cryst. 10:778 (1957).
57. N.G. Vannerberg. Arkiv Kemi 13:29 (1958).
58. A. V. Vedeneev, L.I. Kazarnovskaya, and I.A. Kazarnovskii. Zh. Fiz. Khim. 26:1808 (1952).
59. L.I. Kazarnovskaya. Zh. Fiz. Khim. 20:1408 (1946).
59a. M.M. Pavlyuchenko and T.I. Popova. Dokl. Akad. Nauk BSSR 7:456 (1963).
60. B.F. Clay and R.A. Sahlin. U. S. Patent 2709129 (1955).
61. T. Stevenson and W. Cavell. U. S. Patent 2823205 (1958).
62. A.N. Volskii. Proceedings of the United Nations International Conference on Peaceful Uses of Atomic Energy, Vol. 28, Geneva (1958), p. 170.
63. W. Schumb et al. Hydrogen Peroxide, Moscow, IL (1958).
64. H.H. Hoekje. U.S. Patent 2805128 (1957).
65. P.O. Stelling, E. Angel, and H. Mattson. Swedish Patent 164381 (1958).
66. I.A. Kazarnovskii. The Eighth Mendeleev Meeting on General and Applied Chemistry, Vol. 1, Moscow, Izd. Akad. Nauk SSSR (1959), p. 17.

67. L.A. Isarov. Chemistry of Peroxide Compounds, Moscow, Izd. Akad. Nauk SSSR (1963), p. 103.
68. S.Z. Makarov and N.K. Grigor'eva. Izv. Akad. Nauk SSSR, Otd. Khim. Nauk (1959), p. 9.
69. S.Z. Makarov and N.K. Grigor'eva. Izv. Akad. Nauk SSSR, Otd. Khim. Nauk (1959), p. 1163.
70. A. Marczewski. Polish Patent 43919 (1960).
71. T.B. Pierce. U.S. Patent 2066015 (1937).
72. E. von Drathen and H. Walther. Chem. Ztg. 65:119 (1941).
72a. N.G. Vannerberg. In: Progress in Inorganic Chemistry, Vol. 4, New York, Interscience Publishers, Inc. (1962), p. 162.
73. S. Abrakhams. Usp. Khim. 27:107 (1958).
74. N.G. Vannerberg. Arkiv Kemi 14:147 (1959).
75. N.G. Vannerberg. Arkiv Kemi 14:125 (1959).
76. I.I. Vol'nov and E.I. Latysheva. Zh. Neorgan. Khim. 2:1696 (1957).
77. S.C. Abrahams and J. Kalnais. Acta Cryst. 7:839 (1954).
78. K. Savithri and S. Ramanchandra Rao. Proc. Indian Acad. Sci. 16A:221 (1942).
79. C. Kroger and W. Janetzke. Z. Anorg. Allgem. Chem. 284:83 (1956).
80. A.V. Vedeneev and S.M. Skuratov. Zh. Fiz. Khim. 25:839 (1951).
81. J.P. Coughlin. Bull. 542 Bureau of Mines, Washington (1954), p. 11.
82. I.I. Vol'nov and A.N. Shatunina. Dokl. Akad. Nauk SSSR 110:87 (1956).
83. V.T. Oza. J. Indian Chem. Soc. 33:875 (1956).
84. J.E. Spice and L. Stavenley. J. Soc. Chem. Ind. 68:314, 348 (1949).
85. R. Hill, L. Sutton, R. Temple, and B. White. Research (London) 3:572 (1950).
85a. L.B. Johnson, Jr., Ind. Eng. Chem. 52:137 (1960).
86. F. Booth. Trans. Faraday Soc. 49:272 (1953).
87. G. Huttig. Monatsh. Chem. 85:976 (1954).
88. V. Hogan and S. Gordon. J. Phys. Chem. 61:1401 (1957).
89. K. Hauff. Reactions in Solid Bodies and on Their Surfaces, Pt. 2, Moscow, IL (1963), p. 183.
90. R. Scholder and W. Klemm. Angew. Chem. 66:461 (1954).
91. V.B. Glushkova and O.K. Keler. Zh. Neorgan. Khim. 2:1001 (1957).
92. V.B. Glushkova. Zh. Neorgan. Khim. 2:2438 (1957).
93. H. Fredenhagen. Z. Anorg. Chem. 242:23 (1939).
94. Phillips Petroleum Co. British Patent 790195 (1958).
95. H.F. Park. U.S. Patent 2664416 (1958).
96. N. Ven Kataraman. Current Sci. 21:9 (1952).
97. E. Costa Novella. An. Real Soc. Esp. Fis. Quim. 54B:61 (1958).
98. A. Brengle and H. Stewart. U.S. Patent 2709134 (1955).
99. J. Nevinson and R. Lincoln. U.S. Patent 2491057 (1949).
100. S.G. Stewart. U.S. Patent 2719130 (1955).
101. L.E. Olson. U.S. Patent 2563481 (1951).
102. W.N. Marshall. U.S. Patents 2573442 (1951); 2583584 (1952).
103. W. Hanford. U.S. Patent 2874142 (1959).
104. Chem. Ztb. 126:1888 (1955).
105. R.H. Heyskell. U.S. Patents 2726943 (1955); 2714061 (1955).
106. R.H. Heyskell. U.S. Patent 2726694 (1955).
107. R.H. Heyskell. U.S. Patent 2899291 (1959).
108. T. Toshima. Japanese Patent 8498 (1955).
109. D. Pearsale. U.S. Patent 2909418 (1959).
110. D.T. Zebree. U.S. Patent 2892695 (1959).
111. M. Yamada and J. Yonezawa. Kogyo Kayaky Kyokaishi 19:118 (1958).
112. W.E. Schulz. U.S. Patent 2882819 (1959).
113. B. Carr. Canadian Patent 494258 (1954).

114. L.B. Johnson, Jr. Ind. Eng. Chem. 52:868 (1960).
115. K. Winnacker and E. Weingarten. Chemische Technologie, Vol. 1, München Hanser Verlag (1950), p. 550.
116. C. Weissenberg. U. S. Patent 2763559 (1956).
117. C.W. Becker. U. S. Patent 2559530 (1951).
118. G. Telefunken. British Patent 805880 (1958).
119. A.A. Sheperd. Nature 170:839 (1952).
120. P. Hangelston and R. Ives. U. S. Patent 2545695 (1951).
121. P. Delzien and A. Claude. U. S. Patent 2677623 (1954).
122. P. Delzien and R. Penon. Le Vide 9:257 (1954).
123. E. Oldal and F. Nasel. Austrian Patent 202235 (1959).
124. O. Weirich. U. S. Patent 2868736 (1959).
125. J. Smith. British Patent 688761 (1953).
126. B. King. British Patent 969023 (1959).
127. M.M. Kushnir. Ukr. Khim. Zh. 27:542 (1961).
128. Chim. Ind. Lombarda. Italian Patent 464533 (1951).
129. P.I. Sadovskii. Zavodskaya Lab. 8:1184 (1939).
130. A.D. Brandt. Heating and Ventilation 43:67 (1946).
131. Yu. V. Karyakin and I.I. Angelov. Pure Chemical Reagents, Moscow, Goskhimizdat (1955).
132. S.Z. Makarov and I.I. Vol'nov. Izv.Akad.Nauk SSSR, Otd. Khim. Nauk (1954), p. 765.
133. N.G. Vannerberg. Arkiv Kemi 14:100 (1959).
134. F. Hinz. U.S. Patent 2091129 (1935).
135. I.I. Vol'nov and E.I. Latysheva. Zh. Prikl. Khim. 31:1597 (1958).
136. S.Z. Makarov and L.V. Ladeinova. Izv. Akad. Nauk SSSR, Otd. Khim. Nauk (1957), p. 3.
137. V.F. Boiko. Scientific Reports of Higher Educational Establishment, Khim. i Khim. Tekhnol. 1:57 (1959).
138. V.F. Boiko. Izv. VUZOV, Khim. i Khim. Tekhnol. 5:351 (1962).
139. S.Z. Makarov and L.V. Ladeinova. Izv. Akad. Nauk SSSR, Otd. Khim. Nauk (1957), p. 139.
140. N.G. Vannerberg. Arkiv Kemi 14:2119 (1959).
140a. D.E. Wilcox and J.A. Bromley. Ind. Eng. Chem. 55:26 (1963).
140b. R.C. Ropp and M.A. Aia. Anal. Chem. 34:1288 (1962).
141. S.Z. Makarov and L.V. Ladeinova. Zh. Neorgan. Khim. 1:2708 (1956).
142. W.S. Wood. U.S. Patent 2563442 (1951).
143. C.A. 42:5787 (1952).
144. R.S. Shelton. U. S. Patent 2436673 (1948).
145. L. Harbisson. U. S. Patent 2582829 (1952).
146. E.L. Warrik. U. S. Patent 2718512 (1955); West German Patent 937256 (1955).
147. H. Hock, H. Knopf, and E. Ernst. Angew. Chem. 71:541 (1959).
148. M.H. Abraham. J. Chem. Soc. (1960), p. 4130.
149. M.H. Abraham. Chem. Ind. 25:750 (1959).
150. N.G. Vannerberg. Arkiv Kemi 10:455 (1957).
151. C. Hoffman and R. Kopp. J. Am. Chem. Soc. 81:3830 (1959).
152. L.V. Ladeinova. Izv. Akad. Nauk SSSR, Otd. Khim. Nauk (1961), p. 12.
153. V.F. Boyko. Izv. VUZOV, Khim. i Khim. Tekhnol. 4:171 (1961).
154. A.G. Davies and J.E. Packer. Chem. Ind. (1958), p. 1177.
155. A.G. Davies and J.E. Packer. J. Chem. Soc. (1959), p. 3164.
156. A. Karnojitzki. Chim. Ind. 85:160 (1961).
157. P. Pierron. Bull. Soc. Chim. (1950), p. 291.
158. N.G. Vannerberg. Arkiv Kemi 13:515 (1959).

Chapter Four

Superoxides of the Alkali and Alkaline Earth Metals

The reactions of alkali metals with oxygen are so extensive that most of the elements can be made to form not only higher oxygen compounds, in which the atomic ratio of oxygen to the metal is equal to one, i.e., peroxides, but also higher oxygen compounds in which the ratio is equal to two (superoxides) and even three (ozonides).

The existence of the superoxides of the alkali metals, with the exception of LiO_2, has been positively established. Superoxides of potassium, rubidium, and cesium are produced mainly via the combustion of these metals with oxygen, or with oxygen-enriched air at atmospheric pressure. They are also formed via the oxidation of the metal—liquid ammonia solutions at temperatures below -50°C. Oxidation of metallic sodium at atmospheric pressure results in formation of sodium peroxide, rather than superoxide. The latter is obtained by oxidation of sodium peroxide at a relatively high temperature and at elevated pressure. The oxidation of sodium dissolved in liquid ammonia at -77°C results in the formation of a superoxide—peroxide mixture.

Unlike the corresponding peroxides, which are colorless and diamagnetic, the superoxides of the alkali metals are colored and are paramagnetic. They react very vigorously with moisture and carbon dioxide, and consequently must be stored in hermetically sealed containers. They readily oxidize organic substances and can cause such substances to burn. At room temperature, the alkali metal superoxides can be caused to react with water to liberate all of their active oxygen and form alkali metal hydroxides.   They react with dilute aqueous acid solutions, liberating superoxide oxygen and forming the corresponding salt and hydrogen peroxide.

Of the superoxides, only KO_2 is produced on an industrial scale.   For example, in 1952, the United States produced at least 300 tons of KO_2. Its price was four dollars per pound [1]. Its application, however, is limited. It is used mainly in air revitalization systems, in self-contained isolating type breathing apparatus, and in Kipp's type oxygen generators [2,3].

The superoxides of the alkali metals can be considered as derivatives of the hydroperoxyl radical, HO_2. For this reason, it is appropriate to discuss certain properties of this radical.

The Hydroperoxyl Radical – HO_2

Since the early 1930's, scientists have postulated a role for the HO_2 radical as an intermediate in such reactions as the decomposition of hydrogen peroxide, the reaction of atomic hydrogen with molecular oxygen, and in the reaction of molecular hydrogen with molecular oxygen at high temperatures and low pressures [4].   The HO_2 radical also plays an important role in combustion processes, and it has been detected as one of the reaction products of CH_4, C_2H_4, C_2H_2, and N_2H_4 with oxygen [5, 6].   The existence of this radical was positively established in the period 1952 to 1956 as a result of mass spectrometry studies conducted by A. Robertson [7], S. Foner and R. Hudson [8], and K. Ingold [9]. The special mass spectrometers used in these studies are described in the references [7, 10, 11].

In 1952, Robertson [7] studied the ionization and dissociation of hydrogen peroxide vapor by means of mass spectrometry and reported the formation of the following ions: $H_2O_2^+$, HO_2^+, O_2^+, OH^+, and O^+. In 1958, Robertson's data were confirmed [12] as a result of mass spectrometric analysis of 96% hydrogen peroxide vapor.   It has been established that the HO_2^+ ion concentration is 10% of the $H_2O_2^+$ ion concentration.

Foner [8] has shown that the HO_2 radical is formed in the course of the reaction of atomic hydrogen and molecular oxygen. In these studies, atomic hydrogen was produced in a Wood's chamber using helium as the carrier gas. Molecular oxygen was mixed with an inert gas (CO_2, Ar, He), and the mixture was injected into the atomic hydrogen–helium stream. A mass spectrometer was used to analyze the reaction products. According to Foner, HO_2 is formed via the reaction

$$H + O_2 + M \rightarrow HO_2 + M$$

where M represents the inert gas.

Robertson [13] also conducted an investigation to determine individual elementary stages in the reaction of atomic hydrogen with molecular oxygen in order to explain the mechanism, kinetics, and energy of these stages using mass spectrometry. He has established that at 0.5 mm Hg pressure, approximately 3–4% of the oxygen is transformed to HO_2 and the yield of the latter increases with increase in pressure and reaction time.

The yield of HO_2 radicals via the reaction of atomic hydrogen with molecular oxygen is quite low (HO_2 concentration is approximately 0.001%). Foner and Hudson [14] have succeeded in obtaining higher yields via the reaction of hydrogen peroxide with the dissociation products of various gases exposed to electrical discharge at low pressures (0.5 mm Hg). In the reaction of atomic oxygen with H_2O_2, they observed an HO_2 concentration of approximately 0.01%, whereas in the reaction of atomic hydrogen with H_2O_2, the HO_2 concentration was approximately 0.1% [15]. The concentration of HO_2 obtained by the reaction of hydrogen peroxide with OH radicals, i.e., $OH + H_2O_2 \rightarrow HO_2 + H_2O + 30$ kcal, was 0.3%. The proportion of $O^{17}O^{16}$ in the HO_2 radicals, as determined by these tests, was approximately 3%. The highest HO_2 concentration (up to 0.4%) resulted from the low-power electrical discharge in H_2O_2 vapor [15].

Attempts by Foner and Hudson [16] to detect HO_2 radicals from the reaction of atomic hydrogen and molecular oxygen at liquid nitrogen temperatures were unsuccessful. If HO_2

radicals are formed at such temperatures, it is possible that they condense on the cool surface of the mass spectrometer inlet system and therefore cannot be detected. According to R. Klein [17], the reaction $H + O_2 \rightarrow HO_2$ takes place only at 20°K, i.e., when the trap is cooled with liquid hydrogen. The energy of activation for this reaction is less than 1 kcal/mole. At 20°K, reactions causing the destruction of the HO_2 radical, i.e., $H + HO_2 \rightarrow H_2O_2$ and $2HO_2 \rightarrow H_2O_2 + O_2$, are very weak. Mass spectrometric studies of the products condensed at 4°K from discharge reactions in water or hydrogen peroxide vapor did not reveal the presence of the HO_2 radical [18]. However, electron paramagnetic resonance has been used to detect the presence of HO_2 radicals in ice which has been subjected to ultraviolet irradiation at 77°K [19], in frozen hydrogen peroxide solutions which have been subjected to ultraviolet irradiation at the temperature of liquid nitrogen [20], in the condensate obtained from a glow discharge in water vapor, and from the reaction of solid ozone and atomic hydrogen at the temperature of liquid nitrogen [21].

Ingold [9] used a mass spectrometric method to determine the radicals formed in the high-temperature reaction of molecular hydrogen with molecular oxygen at low pressures. The reaction products included substances with masses 32 (O_2), 33 (HO_2), and 34 (H_2O_2), as well as the OH radical in significant concentration. Mass spectrometric analysis of the negative ions formed in a mixture of water vapor and hydrogen peroxide vapor as a result of electron impact [22] also revealed the presence of HO_2^- ion.

In the hydroperoxyl radical, a three-electron bond is formed between the oxygen atoms. For this reason, the compound should be paramagnetic [23]. Since the HO_2 radical has an odd number of electrons, 13, it cannot have a linear structure [24]. In the ground state, the angle on top of the HO_2 radical is significantly less than 180°. The lifetime of the excited HO_2 radical is, according to Robertson [13], $3 \cdot 10^{-12}$ sec. This value has been confirmed by estimating the recombination rate of HO_2 radicals in the gaseous phase via the reaction

$$2HO_2 \rightarrow H_2O_2 + O$$

The rate constant for the recombination reaction is $3 \cdot 10^{-12}$ ml/mole-sec [25]. The ionization potential of the HO_2 free radical is estimated by Foner [15] as 11.53 ± 0.02 eV, which is 0.7 eV less than the value reported by Robertson [7]. The dissociation energy of the $H-O_2$ bond at 25°C is estimated as 47.1 ± 2 kcal/mole [15], and the dissociation energy of the $HO-O$ bond is estimated as 63.2 ± 4.5 kcal/mole [26]. $\Delta H°_{298}$ for HO_2 is 5.0 ± 2 kcal/mole [15]; $S°_{298}$ is 52.96 eu; and $C_{p}°_{298}$ is 8.34 kcal/mole-deg [27].

The formation of the HO_2 radical in the liquid phase, for example, as an intermediate in the dissociation of hydrogen peroxide by radiation, is described in detail in references [23, 28, 29]. In recent years a number of interesting reports have been published pertaining to the formation of the HO_2 radical in a liquid phase. Of particular interest is a report on the detection of the HO_2^+ ion based on the interpretation of the absorption spectra of aqueous solutions of hydrogen peroxide in the ultraviolet region up to 2100 A [30]. Also of considerable interest is a report on the use of electron paramagnetic resonance to detect the presence of HO_2 radicals in aqueous hydrogen peroxide solutions at room temperature [31]. The enthalpy of solution of the HO_2 radical has been estimated as -8 ± 2.5 kcal/mole and the enthalpy of formation of HO_2 in aqueous solution has been estimated as -3 ± 5 kcal/mole [26]. Reports discussing the DO_2 radical [26, 32] and the TO_2 radical [33] have also been published.

Lithium Superoxide – LiO_2

The existence of lithium superoxide has been postulated [34-36] on basis of similarities in the absorption spectra of liquid ammonia solutions of metallic lithium, sodium, and potassium oxidized at -77°C. Following oxidation, all the solutions are yellow, and the absorption spectra of the three solutions show a peak at 38 mμ. Sodium and potassium superoxide can be recovered from the liquid ammonia solution upon evaporation of the liquid ammonia. However, attempts to recover lithium superoxide in a similar manner have been unsuccessful.   Upon heating, the oxidized lithium–liquid ammonia solution loses its color at -33°C, and the recovered product contains a mixture of Li_2O and Li_2O_2. Another indirect proof of possible LiO_2 formation is based on the chemical analysis of the reaction products obtained from the reaction of lithium nitrate with sodium superoxide at -30°C in liquid ammonia.   The reaction proceeds according to the following:

$$2LiNO_3 + 2NaO_2 \rightarrow Li_2O_2 + O_2 + 2NaNO_3$$

The oxygen evolved in this reaction is attributed to the dissociation of unstable LiO_2 which is formed as an intermediate compound, rather than to the dissociation of the sodium superoxide [37].   A more convincing proof for the existence of LiO_2 comes from the analysis of the solid products obtained as a result of the vacuum dissociation of $Li_2O_2 \cdot 2H_2O_2$. It has been reported that the product obtained following the dissociation of this compound at 100 to 120°C and 10 mm Hg pressure is a mixture of Li_2O_2, LiOH, and 7 to 9 wt.% LiO_2 [38, 39]. Attempts by the author of this book to synthesize LiO_2 via the reaction $Li_2O_2 + O_2 \rightarrow 2LiO_2$ at 200°C and 7000 atm oxygen pressure were not successful.

N. Vannerberg [40] attributes the instability of lithium superoxide to the small radius of the lithium ion (0.60 A). In order to achieve a stable superoxide structure, it appears that the radius of the cation should be in the range 0.66-1.15 A.

Sodium Superoxide — NaO_2

Production Methods.   In 1949, I.A. Kazarnovskii [41] reported that as early as 1936 he had succeeded in synthesizing sodium superoxide via the reaction

$$Na_2O_2 + O_2 \rightarrow 2NaO_2$$

The feasibility of NaO_2 formation was mentioned at the end of the last century by E. B. Schoene [42] on the basis of his studies of the dissociation of $Na_2O_2 \cdot 2H_2O_2$; by R. de Forcrand [43] on the basis of thermodynamic considerations; and later by F. Haber [44] on the basis of his sodium oxidation kinetic studies. Schoene's method has been the subject of detailed studies in recent years [45]. It has been demonstrated that $Na_2O_2 \cdot 2H_2O_2$ can be caused to disproportionate upon vacuum desiccation at 70-120°C and 10 mm Hg to yield a mixture of Na_2O_2, NaOH, and NaO_2, containing up to 30 wt.% of the latter. In a vacuum at 40°C, only 20% of $Na_2O_2 \cdot 2H_2O_2$ is transformed to NaO_2 [46]. At 60°C and the same residual pressure, the product obtained contains 25 wt.% NaO_2, and the yield is less than 50%. It is of interest to note that when $Na_2O_2 \cdot 2H_2O_2$ was treated in dried air at 80°C at 1 atm pressure and the reaction was stopped before the superoxide content reached 10%, yields as high as 90% were obtained [47].

Since 1948, J. Kleinberg, along with his associates, has carried out a number of investigations on the synthesis, analysis, and study of certain properties of sodium superoxide [48-52]. According to the published data, the rapid oxidation of metallic sodium in liquid ammonia at -77°C results in the formation of a product with the empirical composition $NaO_{1.67}$, which corresponds to a mixture of 1 mole sodium peroxide and 4 moles sodium superoxide. The solubility of sodium superoxide in liquid ammonia at -50 to -33°C is 0.006 ± 0.002 g/100 ml NH_3, or $1.4 \cdot 10^{-3}$ g/mole NH_3 [53]. Recrystallization from liquid ammonia has been used to produce NaO_2 crystals for x-ray structural analysis. Probably the most efficient method for obtaining NaO_2 is the one suggested by Kazarnovskii [41].

Kleinberg et al. have obtained a product containing approximately 96% NaO_2 by oxidizing sodium peroxide at 450-475°C and 280 atm of oxygen. Information concerning the reaction time for the completion of this reaction is contradictory. In reference [51] it is stated that the reaction is complete in 100 hr, whereas in reference [52] the reaction time is reported as 6 hr.   Improvements in this method of synthesis are described in references [3, 54, 55].   It has been established that the formation of sodium superoxide takes place at temperatures not less than 200°C, and pressure not less than 105 atm. At 400°C and 122 atm, it is possible to obtain almost 100% pure NaO$_2$ [3, 46].   The amount of NaO$_2$ formed as a function of the O$_2$ pressure over Na$_2$O$_2$ for temperatures in the range of 300 to 400°C is shown in Fig. 15 [55]. The reaction can be accelerated by mixing metallic oxides, for example, CdO and TiO$_2$, with the sodium peroxide.   The highest conversion of sodium peroxide to superoxide is achieved by using sodium peroxide obtained by oxidizing molten and sprayed sodium, since the peroxide obtained in this manner has the highest degree of dispersion [3]. According to patent information, sodium peroxide can be produced by direct oxidation of metallic sodium at temperatures above 200°C and at a pressure of 175 atm. However, it has not been possible, as yet, to select proper materials for use as reactors which would be sufficiently stable to allow this reaction to be carried out [35, 55]. It is possible that the conditions reported for obtaining NaO$_2$ by direct oxidation of metallic sodium are incorrect. In a recently

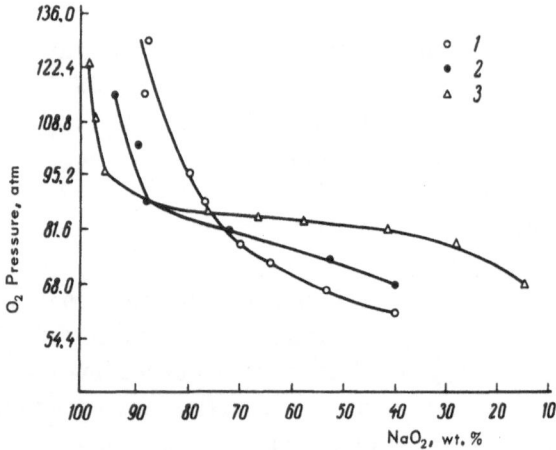

Fig. 15. NaO$_2$ content in Na$_2$O$_2$ as a function of oxygen pressure [55]; (1) 300°C; (2) 350°C; (3) 400°C.

published article, it was reported [56] that the synthesis of NaO_2 was achieved by the reaction of oxygen and an evaporated film of sodium in the presure range of 10^{-3} to 10^{-1} mm Hg and 0°C.   It is interesting to note that the sodium cloud created for determining a rocket's trajectory by dispersing metallic sodium into high layers of the atmosphere basically consists of sodium superoxide vapor [57].

According to a 1954 patent [55], sodium superoxide can be produced by oxidizing sodium oxide at 300-500°C and 130 atm oxygen pressure.   In 1959, a patent was issued for the production of sodium superoxide via the oxidation of metallic sodium dissolved in organic bases, especially pyridine [58]. Evaluation of this method by the author of this book indicated that the main oxidation product is sodium peroxide, rather than sodium superoxide.   The most promising method for producing sodium superoxide via reactions involving the use of organic substances is probably the one suggested by A. LeBerre [59].   In this method, NaO_2 is produced as a result of the self-oxidation of sodium benzophenone.   The sodium benzophenone is prepared via the reaction of metallic sodium with benzophenone dissolved in tetrahydrofuran.   The reaction scheme is as follows:

$$\begin{array}{l}C_6H_5\\C_6H_5\end{array}\!\!>\!\!C=O \xrightarrow{+Na} \begin{array}{l}C_6H_5\\C_6H_5\end{array}\!\!>\!\!C.\!\!\nearrow^{ONa} \xrightarrow{+O_2}$$

$$\begin{array}{l}C_6H_5\\C_6H_5\end{array}\!\!>\!\!C\!\!<\!\!\begin{array}{l}ONa\\OO.\end{array} \rightarrow \begin{array}{l}C_6H_5\\C_6H_5\end{array}\!\!>\!\!C=O + NaO_2$$

The recovered sodium superoxide is 95-98% pure. The yield of the superoxide with respect to sodium is 85%, and with respect to oxygen is 90%.   It is of interest to note that NaO_2 remains dissolved in the tetrahydrofuran and it is separated only after another solvent has been added, as, for example, ligroin.   The dried NaO_2 does not dissolve in tetrahydrofuran.

Properties of Sodium Superoxide. With respect to active oxygen content (43.6 wt.%), sodium superoxide exceeds all known solid peroxide compounds, except the relatively unstable ozonides of sodium (NaO_3) and potassium (KO_3). It has a yellow color and is paramagnetic. Its effective magnetic moment is 2.07 Bohr magnetons [51], corresponding to a structure possessing one unpaired electron, which is in accordance with the electron formula Na^+ [$:O\cdot\negthickspace\cdot O:$]$^-$, which has been confirmed by paramagnetic resonance studies [60]. In mixtures of sodium peroxide and sodium superoxide, the NaO_2 paramagnetism shows a linear increase with increase in NaO_2 content. Measurement of the magnetic susceptibility of such mixtures can thus be used to determine the amount of NaO_2 present with an accuracy of ± 3% [51]. The O_2^- ion is homopolar, and consequently NaO_2 is inactive in the infrared region [61].

Sodium superoxide is stable up to 65°C when stored in an hermetically sealed container. It begins to disintegrate at 100°C with the liberation of oxygen [55]. According to reference [62], its thermal disintegration begins at 100 to 120°C with the formation of a series of solid solutions. A solid solution with a composition $Na_2O_{3.6}$ disintegrates to sodium peroxide at 250°C. At 400°C, the solid sodium peroxide begins to slowly decompose. Sodium peroxide then begins to melt, and at 540°C it violently disintegrates with the formation of sodium oxide.

Three crystal modifications of NaO_2 are known. At 25°C, sodium superoxide has a face-centered cubic lattice of the pyrite type with a = 5.49 ± 0.005 A [63-65]. This modification is usually indicated by the formula NaO_2(I) and is stable to -43°C. In the NaO_2(I) lattice, each sodium ion is surrounded by six oxygen atoms. The density of NaO_2(I) is 2.21 g/cm^3. The specific heat capacity of NaO_2(I), $C_p°_{298}$, is 17.24 cal/mole-deg, and $S°_{298}$ is 27.7 ± 0.3 eu [66]. The existence of the low temperature modifications has been established by magnetic measurements [67], by thermal analysis [68], and by x-ray structural analysis [63]. The structural changes of NaO_2 at different temperatures are accompanied by sudden changes in certain of its physical properties. For example, the magnetic susceptibility–temperature curve shows a distinct peak at -80°C [67]. The differential cooling curves show two exothermal effects at -43 and -80°C [68]. Below -50°C, NaO_2 has a cubic lattice, with a = 5.46 A.   Its density is 2.24 g/cm^3. This modification is usually designated as NaO_2(II). Below -77°C, the sodium superoxide lattice is rhombic, of the marcasite type, with constants a = 4.26 A, b = 5.54 A, and c = 3.44 A. Its density is 2.25 g/cm^3 [63].   This modification is usually designated as NaO_2(III). These data can be closely correlated with heat capacity data [66].   The heat capacity–temperature curve shows two distinct peaks at –49.9 and -76.7°C. By cooling sodium superoxide to the temperature of liquid air, it loses its color [69].   The NaO_2 lattice energy is estimated as 190.9 kcal/mole [70].   The magnetic susceptibility of NaO_2 is χ_m = 33.0 $\cdot$ 10^{-6} [50, 67].

For the reaction $\frac{1}{2}Na$ + $\frac{1}{2}O_2$ → $\frac{1}{2}NaO_2$, $\Delta H°_{298}$ = -31.05 ± 0.35 kcal [71], $\Delta F°_{298}$ = -25.9 kcal, and $\Delta S°_{298}$ = -16.8 eu [72]. Reference [73] presents values for ΔH and ΔF for the temperature range of 298 to 1000°K. For the reaction Na_2O_2 + O_2 → $2NaO_2$, $\Delta H°_{298}$ = 400 cal, and $\Delta S°_{298}$ = -16.2 eu [74].

At room temperature, sodium superoxide reacts with water to completely liberate its active oxygen in 100 sec [75]. The heat of reaction of NaO_2 with H_2O is 15.9 ± 0.7 kcal/mole [71]. The reaction of NaO_2 with water vapor, at room temperature, is accompanied by the liberation of all active oxygen from the superoxide and by the formation of sodium hydroxide monohydrate.   At lower temperatures, 0, -5, -10°C, it liberates only superoxide oxygen with the formation of sodium peroxide crystallohydrates [76].

The quantity of peroxide and superoxide active oxygen in superoxides of sodium and other alkali metals can be determined gasometrically, via the reaction of the superoxide with a mixture of 2M HCl and 1M $FeCl_3$ solution [77],or with a 0.5% aqueous solution of copper sulfate [78]. Determination of the amount of peroxide and superoxide oxygen in mixtures can be accomplished using two separate samples or one sample. In the first case, one sample is used to determine the total active oxygen content by one of the methods described above [77, 78]. A second sample is used for determination of peroxide oxygen only. This may be done by titration of the sample with 0.1N potassium permanganate solution at a low temperature in an acid medium [78]. The difference of the two determined values is equal to the amount of superoxide oxygen. In the second case, approximately 0.2 g superoxide is mixed with 5 ml diethylphthalate. To this is added 10 ml of a mixture containing ice-cold acetic acid (8 ml) and diethylphthalate (2 ml). In this process, only superoxide oxygen is liberated and can be measured gasometrically. To determine the peroxide oxygen, 15 ml of a mixture of 1M HCl and 3M $FeCl_3$ is added to the acetate solution and the volume of liberated peroxide oxygen is measured [77]. The peroxide oxygen may also be determined by titration of an aliquot of the acetate solution with permanganate. Up to 1 wt.% NaO_2 mixed with sodium peroxide can be determined by measuring the oxygen liberated by the reaction of 5 g of the sample with a mixture consisting of 1 ml water and 9 ml ethyl alcohol [78b].

The reaction of sodium superoxide with carbon dioxide in the presence of water vapor at 25°C results in the formation of sodium carbonate and the liberation of all of the active oxygen. At temperatures below 10°C, only the superoxide oxygen is liberated and sodium peroxydicarbonate, $Na_2C_2O_6$, is formed [78]. Sodium superoxide reacts with dry carbon monoxide at temperatures above 100°C, and in the presence of water vapor it reacts at 95°C with the formation of sodium carbonate [79]. Even when heated, NaO_2 cannot be made to react with dry sodium carbonate or with anhydrous sodium bicarbonate [80-82]. It begins to react with the dissociation products of bicarbonate, i.e., with H_2O and CO_2, at 100°C. At -30°C, sodium superoxide reacts with liquid NH_3 solutions of the nitrates of calcium, strontium, magnesium, cadmium, and lithium to form the peroxides of these metals [37].

Potassium Superoxide – KO_2

Production Methods. Potassium superoxide was discovered in the last century [83]. It was obtained in a highly pure state by oxidizing potassium metal dissolved in liquid ammonia at -50°C [51, 84]. The essential condition for the production of a

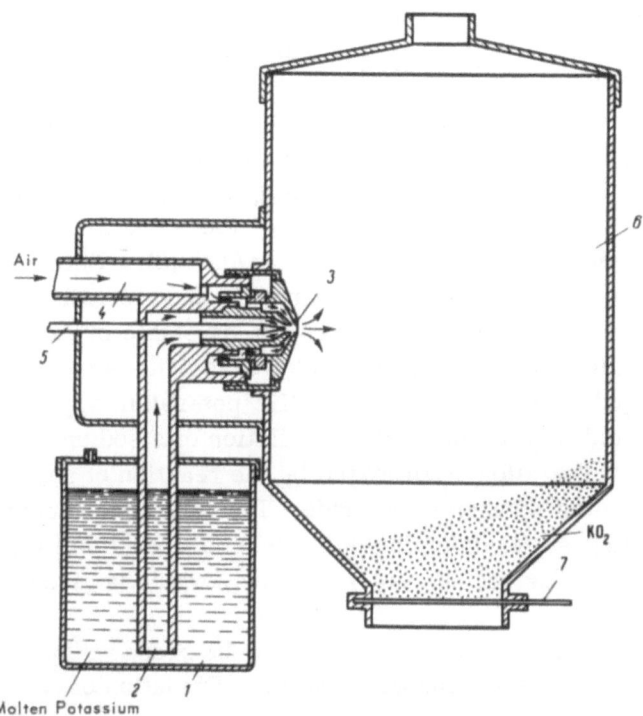

Fig. 16. Industrial unit for KO_2 production [86]. (1) Container with molten metallic potassium; (2) pipe for supplying molten potassium to the sprayer; (3) sprayer; (4) pipe for supplying air to the nozzle; (5) needle for controlling the supply of air and molten potassium mixture to the nozzle; (6) cylindrical KO_2 container; (7) KO_2 outlet.

pure product is the absence of moisture in the ammonia. The presence of moisture results in the formation of KOH, KNH_2, and KNO_3 mixtures. For example, due to the presence of 0.05% water in ammonia, the recovered product contained 20.0% KOH, 28.4% KNH_2, and 6.2% KNO_3 [85].

On an industrial scale, KO_2 is produced by the oxidation of dispersed liquid potassium with air enriched with oxygen (from 13 to 35 vol.%). The oxygen-enriched air is heated to 75-79°C under an excess pressure of 0.13 atm. The temperature in the reaction zone is approximately 300°C. The amount of air used is 5 to 15 times greater than the theoretical amount required for the oxidation reaction [86, 87]. Figure 16 shows a unit that is used for the industrial production of KO_2 [86].

In order to prevent the loss of the more finely-divided potassium superoxide powder, a conventional Cottrell precipitator is used to collect the KO_2 dust [88]. The production of KO_2 in the United States by the Mine Safety Appliance Company is approximately 3 tons/day [3]. Pure potassium superoxide, containing approximately 2 wt.% KOH and K_2CO_3 and free from metallic potassium, is stable in storage [88a]. According to a 1949 patent [89], potassium superoxide can be obtained by oxidizing the metallic potassium vapor which is liberated as a result of the distillation of a sodium–potassium alloy. The alloy is prepared by the reaction of sodium vapor and molten potassium chloride. The vapor phase combination of a KNa alloy containing 60% potassium results in the formation of a mixture of Na_2O_2 and KO_2, which is given the name "MOX" [3].

Potassium superoxide can also be prepared by the oxidation of potassium amalgam [90, 91]. Oxidation of other potassium compounds also results in the formation of KO_2. According to reference [92], the oxidation of K_2O to KO_2 (Fig. 17)

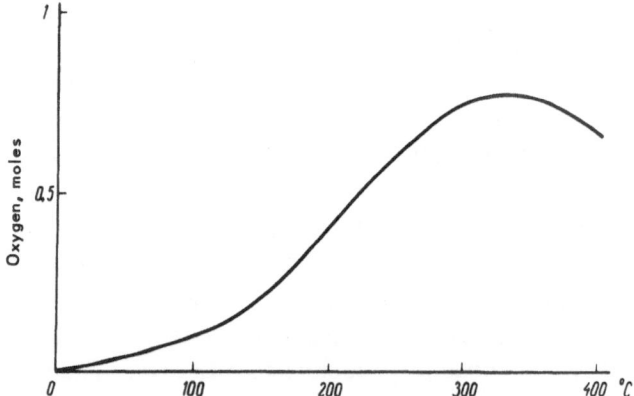

Fig. 17. The amount of oxygen uptake as a function of temperature for the reaction $K_2O + 1.5O_2 \rightarrow 2KO_2$ [92].

proceeds intensively, starting at 200°C, and becomes complete at 350°C. The oxidation of KNH_2 [93] also results in the formation of KO_2. It has been confirmed that potassium superoxide can be synthesized by the vacuum desiccation of $K_2O_2 \cdot 2H_2O_2$ [94, 95]. KO_2 can also be synthesized via the oxidation of potassium alcoholates dissolved in organic solvents [96]. For example, the passage of oxygen for 30 min through a mixture of 25 millimoles (4.605 g) of benzhydrol and 50 millimoles (5.610 g) of potassium tertiary butylate in anhydrous benzene produces a yellow deposit which analyzes as 89.5 wt.% KO_2, 3-4% organic substances, and 6.5-7.5% KOH [97]. Tetrahydrofuran can be used as the solvent in place of benzene [98].

Of considerable interest are the studies which have been reported on the synthesis of KO_2 via the reaction

$$2KOH + 1.5O_2 \rightarrow 2KO_2 + H_2O$$

It has been shown that at 210°C and 1 atm O_2 pressure, the oxidation of KOH yields a product containing 20% KO_2 [99]. At 510°C and 1 atm O_2 pressure, a product containing approximately 32% KO_2 was recovered [84], and after 5 hr at 375°C and 100 atm O_2 pressure, the product contained 70% KO_2 [100].

The author of this book has observed the formation of KO_2 via the reaction of KOH and atomic oxygen, generated by means of a glow discharge. The vapor absorption of 96-98% hydrogen peroxide by potassium hydroxide at 55°C and 10-20 mm Hg pressure is reported to produce a product containing up to 35 wt.% KO_2 [101]. The formation of a colored compound was observed by passing ozone through concentrated (40 wt.%) KOH solution at -40°C [102]; however, the compound has been found to be an ozonide, rather than potassium superoxide.

Properties of Potassium Superoxide. Potassium superoxide is a yellow, paramagnetic solid. Its magnetic moment is 2.04 Bohr magnetons [67,103]. It is a more thermally stable compound than sodium superoxide. According to Rode [104], it begins to decompose at 145°C. In 1932, Blumenthal [105] reported the melting point of KO_2 to be 380°C, and this value has found its way into most handbooks as the accepted value for the melting point of KO_2. It is now felt that this value is too low. In fact, even the value of 440°C, reported by de Forcrand [105a] in 1914, is also thought to be too low. According to a 1963 reference, KO_2 does not even begin to decompose to K_2O until it is heated to 425°C [106].

Potassium superoxide can exist in one of four crystal modifications. The α-KO_2 modification is stable in the temperature range -75 to 100°C. It has a tetragonal lattice with a = 5.704 ± 0.005 A, c = 6.699 ± 0.005 A [107, 108, 109], and Z = 4 [106]; its density is 2.158 ± 0.001 g/cm^3 [110]. The values for the refractive index and molecular refraction of α-KO_2 are 1.450 and 8.85, respectively [110]. In the KO_2 lattice, each cation is surrounded by ten oxygen atoms, as in the BaO_2 lattice, rather than by six oxygen atoms, as in NaO_2. Similar to the BaO_2 lattice, the interatomic distances of the metal and oxygen atoms differ in the respect that in the plane parallel to the c axis, the K—O distance is 2.71 A, whereas in the plane perpendicular to the c axis, the distance is 2.92 A [109]. Figure 18 shows the tetragonal coordination of the superoxide ion, O_2^-, in the KO_2 lattice [111].

Above 100°C, α-KO_2 is transformed to β-KO_2, which is a face-centered cubic crystal with a = 6.09 ± 0.01 A. β-KO_2 is isomorphous with NaO_2(I) [65, 112]. Low temperature heat capacity studies [66], as well as differential thermal analysis studies [68], support the existence of two low temperature phases of KO_2 with phase transition temperatures of -75 and -120°C.

For the reaction $\frac{1}{2}K + \frac{1}{2}O_2 \rightarrow \frac{1}{2}KO_2$, $\Delta H°_{298} = -33.8 \pm 0.4$ kcal [71], $\Delta F°_{298} = -28.7$ kcal [71], and $\Delta S°_{298} = -18.2$ eu [71]. The heat capacity of KO_2, $C_p°_{298}$, is 18.53 kcal/mole–deg, and the entropy, $S°_{298}$, is 27.9 ± 0.6 eu [67].

The heat of reaction of KO_2 with water [71] is +13.2 ± 0.8 kcal/mole, which is approximately three times less, per unit of weight, than the heat associated with the reaction of sodium peroxide with water. In the reaction with water at room temperature, potassium superoxide completely liberates its

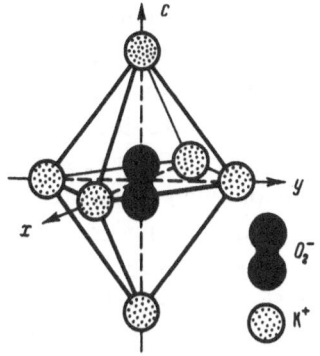

Fig. 18. Tetragonal coordination of the superoxide O_2^- ion in the KO_2 lattice [111].

active oxygen in 100 sec [75]. On a unit weight basis, the amount of O_2 obtainable from the KO_2-H_2O reaction is approximately 1.5 times more than the amount obtainable from the $Na_2O_2-H_2O$ reaction. For this reason, potassium superoxide has been recommended for use in special oxygen generators of the Kipp type used in underwater metal cutting [2]. A mixture of potassium superoxide and sodium peroxide containing 26.8 wt.% active oxygen has been used to supply oxygen for field hospitals [112a]. Upon reaction with cold water, these mixtures liberate approximately 89% of their active oxygen. Oxygen generators built on this principle, with capacities of from 23 to 680 liters of oxygen, were developed in the United States. The largest generator of this type can be used to support 20 patients.

In the reaction of potassium superoxide with water vapor at 10°C and lower, only the superoxide oxygen is liberated, and potassium peroxide hydrates, $K_2O_2 \cdot nH_2O$, are formed. At 19°C and higher, all active oxygen is liberated with the formation of potassium hydroxide and its hydrates [113].

The reaction of potassium superoxide with water vapor and carbon dioxide results in the liberation of oxygen and the absorption of an equivalent amount of CO_2. For this reason, it finds application in self-contained types of respiration apparatus [2, 3]. For example, members of the Swiss expedition which conquered Mount Everest in 1952 made use of an American-made apparatus weighing 2.25 kg, with which they successfully generated oxygen for 45 min to altitudes of approximately 3000 m [114]. In the United States, fire-fighting and mine-rescue crews are equipped with respiration units using potassium superoxide [3, 23]. The efficiency of potassium superoxide for air regeneration purposes depends on the design of the respiration apparatus [115-117]. At best, the use of such apparatus under physical stress is limited to 1 hr, allowing the consumption of up to 80% of the available active oxygen. For use in canisters for respiratory equipment, the fluffy KO_2 powder is pressed and then granulated and screened to 2 to 3 mesh [3], resulting in a charge with a density of approximately 0.66 g/cm^3 [88a]. Various attempts have been made to produce superoxide canisters which have less breathing resistance and operate at a lower temperature than the originally developed canisters. A 1957 patent [118] describes a canister design which is intended to achieve these characteristics. In that invention, KO_2 particles vary in size from 0.5 to 2.1 mm and are arranged in layers separated by screens. Particles in the range of 0.95 to 2.1 mm compacted under a pressure of 122 atm form one layer, and particles in the range of 0.5 to 0.95 mm compacted under a pressure of 91 atm form another layer. In order to insure an immediate supply of oxygen, a potassium chlorate candle is made a part of some KO_2 canisters [118a].

In 1960, the United States press published information on experiments that were carried out on the use of potassium superoxide for the purification of air in spaceship cabins [88a]. It was reported that in a cabin with a volume of 5.943 m^3, 850 g KO_2 provided enough oxygen to support one man for 12 hr, and the efficiency with respect to the utilization of the KO_2 was as high as 99.8%. Potassium superoxide is effective as an air regeneration material, not only because of its ability to remove CO_2 from the breathing atmosphere, but because it also removes such toxic contaminants as indol, skatol, and hydrogen sulfide [118b]. It is also effective in the removal of airborne bacteria.

Potassium superoxide does not react with dry carbon dioxide. The heat of reaction for

$$2KO_2 + CO_2 \rightarrow K_2CO_3 + 1.5O_2$$

has been estimated to be 43.1 kcal [88a].

In the reaction of potassium superoxide with moisture-containing carbon dioxide at 10°C or less, only the superoxide oxygen is liberated and the peroxydicarbonate $K_2C_2O_6$ is formed. At 50°C or above, all active oxygen is liberated with the formation of potassium carbonate and bicarbonate [113]. Potassium carbonate is also formed in the reaction of potassium superoxide with dry carbon monoxide at 95°C [78]. According to reference [118b], the reaction of moisture-containing carbon dioxide with potassium superoxide can be caused to proceed with the formation of $KHCO_3$ at temperatures below 60°C. Under these conditions, the ratio of absorbed CO_2 to liberated oxygen is more than 0.67. In the reaction of potassium superoxide with a mixture of CO_2 and water vapor in a ratio of 2:1, the CO_2/O_2 value is approximately 0.82. The rate at which the gas mixture is supplied to the KO_2 does not significantly affect the reaction.

Potassium superoxide can be caused to react with ozone to form potassium ozonide, KO_3 [119]. The reaction of calcium, strontium, magnesium, and zinc nitrates with potassium superoxide, in molar ratios of 4:1 and 2:1, in liquid ammonia at -30°C results in the formation of the respective metal peroxides. Cadmium nitrate and lithium nitrate do not react with potassium superoxide under the above-mentioned conditions [37], whereas in the case of barium nitrate, the compound $K_2Ba(O_2)_2O_2$ is formed [120]. The reaction of potassium superoxide with mineral acids proceeds as follows:

$$2KO_2 + 2HCl \rightarrow 2KCl + H_2O_2 + O_2$$

The reaction will not take place under completely anhydrous conditions or at temperatures below -40°C [121]. Potassium superoxide reacts with liquid SO_3 dissolved in SO_2 at -20 to -15°C with the formation of the potassium salt of monoperoxy-tetrasulfuric acid, $K_2S_4O_{14}$, via the reaction [122]

$$2KO_2 + 4SO_3 \rightarrow K_2S_4O_{14} + O_2$$

With trimethylchlorosilane, potassium superoxide reacts via the reaction

$$2KO_2 + 2(CH_3)_3SiCl \rightarrow 2KCl + (CH_3)_3SiOSi(CH_3)_3 + 1.5O_2$$

to form hexamethyldisiloxane [121]. In the presence of moisture, KO_2 reacts partially with the methyl and ethyl ethers of chlorosulfonic acid to form KCl and K_2SO_4 and the respective alcohol. However, in the absence of moisture, no reaction takes place, even at temperatures up to 100°C. Under anhydrous conditions, KO_2 does not react with benzoyl chloride or with triphenylchloromethane. However, in the presence of moisture, reaction of KO_2 with these substances will take place, leading to the formation of benzoyl peroxide and triphenylmethyl peroxide [121].

Until recently, the compound K_2O_3 was thought to exist [83]. However, infrared studies of oxides analyzing as K_2O_3 do not reveal the presence of the O_3^{2-} ion [123]. In addition, vacuum thermal decomposition studies of KO_2 [110] revealed that the oxygen pressure over the sample remains constant until the K_2O_2 composition is reached. These two experimental facts indicate that K_2O_3 does not exist. It is probable that in the oxidation process of metallic potassium, a secondary reaction of the metal suboxide and KO_2 occurs, by which mixtures of KO_2 and K_2O_2 are formed which are accepted as individual compounds. This possibility was first proposed by D. I. Mendeleev [124]. In one of the latest reviews [125], these mixtures are assigned the formula $K_4^+(O_2^-)_2O_2^{2-}$ and the name tetrapotassium peroxide disuperoxide, in spite of the fact that x-ray structural investigations do not support the existence of such a compound [126].

Rubidium Superoxide — RbO_2

The methods for the synthesis of rubidium superoxide are similar to those used for the synthesis of potassium superoxide. A product containing 98 wt.% RbO_2 has been obtained by the oxidation of a metallic rubidium–liquid ammonia solution at -50°C [108]. RbO_2, 99.1% pure, has been obtained by the direct oxidation of small amounts (0.2 to 0.5 g) of the metal at elevated temperatures [127]. The metallic rubidium was placed in a reactor filled with nitrogen and its oxidation occurred as the nitrogen was gradually replaced with oxygen. Less pure products have been obtained by oxidizing RbOH at 410°C and 1 atm pressure and as a result of the vacuum desiccation of $Rb_2O_2 \cdot 4H_2O_2$ at 56°C. In the first method, the product contained 56.5 wt.% RbO_2 [84], whereas the second method yielded a product containing 76.3% RbO_2 [128].

Rubidium superoxide is a yellow, paramagnetic solid. Its magnetic moment is 1.89 Bohr magnetons [108]. The unit cell of rubidium superoxide is tetragonal with constants a = 6.00 A and c = 7.03 A [108]; its density is 3.06 g/cm^3 [108]. The thermodynamic properties of RbO_2 given in Table XI are taken from reference [73].

The thermal decomposition of RbO_2 has been studied in the temperature range from 280 to 360°C [127]. For the reaction $RbO_2 \rightarrow \frac{1}{2}Rb_2O_2$(solid) $+ \frac{1}{2}O_2$(gas) at 320°C, $\Delta H° = 9.30$ kcal, $\Delta S° = 7.39$ eu, $K_p = P_{O_2}^{1/2} = 2.10 \cdot 10^{-2}$, and $\Delta F° = 4.55$ kcal. For the reaction Rb_2O_2(solid) $\rightarrow Rb_2O$(solid) $+ \frac{1}{2}O_2$(gas) at 330°C, $\Delta H° = 6.85$ kcal, $\Delta S° = 2.13$ eu, $K_p = P_{O_2}^{1/2} = 8.78 \cdot 10^{-3}$, and $\Delta F° = 5.58$ kcal. In the course of these studies on the thermal decomposition of RbO_2, the investigators did not detect the formation of a Rb_2O_3 oxide or of solid solutions of RbO_2 and Rb_2O_2 [127]. This is additional proof for discounting the existence of such alkali metal compounds. The existence of the rubidium and cesium compounds, $Rb_4^+(O_2^-)_2O_2^{2-}$ and $Cs_4^+(O_2^-)_2O_2^{2-}$, has not as yet been ruled out [126]. It has been reported that these compounds have a cubic lattice with a = 9.30 A for the rubidium compound and a = 9.86 A for the cesium compound.

Cesium Superoxide – CsO$_2$

Production Methods. Cesium superoxide, as potassium and rubidium superoxides, can be obtained by any one of the following six methods: (a) oxidation of the metal dissolved in liquid ammonia, (b) direct oxidation of the metal, (c) oxidation of cesium amalgam, (d) oxidation of cesium oxide, (e) oxidation of cesium hydroxide, and (f) the disproportionation of cesium peroxide peroxyhydrate.

Cesium superoxide was first prepared by means of method (b) by E. Rengade in 1906 [129-131]. In reference [129], in which Rengade first reported the discovery of cesium super-oxide via the oxidation of metallic cesium dissolved in liquid ammonia, the properties of the compound are not described, nor are any analytical data presented. It is simply stated that upon the passage of excess oxygen through liquid ammonia containing dissolved cesium at -50 to -70°C the compound formed is CsO_2, i.e., the superoxide rather than the peroxide. In reference [130], it is noted that CsO_2 can be prepared from the reaction of oxygen with Cs_2O at 150°C; again the properties of the material and analytical data are not presented. More detailed information is given in references [131, 132], in which the preparation of CsO_2 via the oxidation of metallic cesium is discussed. Reference [132] describes a laboratory method for the production of large quantities of CsO_2 of 97% purity.

The other methods for the synthesis of cesium super-oxide do not yield products of such high purity. For example, the product obtained as a result of the oxidation of cesium amalgam with dry oxygen [90] had the empirical composition $Cs_2O_{3.6}$. In 1910, de Forcrand [43] concluded, on the basis of thermodynamic calculations, that cesium superoxide can be synthesized by the oxidation of CsOH at approximately 500°C, provided the water formed in the reaction

$$2CsOH + 1.5O_2 \rightarrow 2CsO_2 + H_2O$$

is removed as it is formed. This method was experimentally demonstrated two years later [100]. The oxidation of CsOH at 100 atm and 350°C yielded a product believed to be a mixture of cesium peroxide and superoxide. Analysis of the mixture on the basis of Cs_2O_2 showed it to contain 95.5% Cs_2O_2. In a later experiment [84], in which the oxidation was carried out under anhydrous conditions at atmospheric pressure and 410°C, the product was positively shown to contain CsO_2, since the analysis based on Cs_2O_2 gave a value of 164.9% Cs_2O_2. Vacuum desiccation of $Cs_2O_2 \cdot xH_2O_2$ at 78-100°C has yielded a mixture consisting of 93.7% CsO_2, 0.28% $CsOH$, and 5.99% water [128].

Properties of Cesium Superoxide.   Cesium superoxide is a yellow solid.   A white modification can be formed via the oxidation of a cesium metal film by liquid oxygen [133]. According to Rengade [130], CsO_2 melts at 515°C in an oxygen atmosphere. From dissociation pressure measurements [134], its melting point was estimated as 432°C, whereas, from differential heating curves, the melting point was established as 450°C [132]. Cesium superoxide begins to decompose at 350°C, and at 400°C its dissociation pressure reaches 7 mm Hg. The determination of its density, measured in toluene at 19°C, yielded a value of 3.77 g/cm^3. A density of 3.68 g/cm^3 was obtained on the basis of measurements in kerosene at 0°C.

In reference [73], the thermodynamic properties of cesium superoxide are given as $\Delta H°_{298}$ = -62.0 ±10 kcal/mole, $\Delta F°_{298}$ = -50.5 ± 11.5 kcal/mole, and $\Delta S°_{298}$ = -18 eu. Studies of the thermal decomposition of CsO_2 have been carried out in the temperature range 280 to 360°C [134a]. CsO_2 does not form solid solutions with Cs_2O_2.

Like all superoxides, CsO_2 contains the O_2^- ion, and is, therefore, paramagnetic. The magnetic susceptibility of CsO_2 at 293°K is χ_g = 9.3 $\cdot$ 10^{-6}, which corresponds to an effective magnetic moment (μ_{eff}) of 1.89 Bohr magnetons [108]. Powder x-ray studies [108] of 98% pure CsO_2 obtained by the oxidation of metallic cesium in liquid ammonia indicated that its basic cell is tetragonal with the constants a = 6.28 A and c = 7.24 A. The x-ray density is 3.80 g/cm^3, which corresponds to the value determined pycnometrically. The number of molecules in the basic cell is four. The lattice energy is estimated [135] as 158.8 kcal/mole.

Cesium superoxide is inert to pure alcohol. It reacts with water via the reaction

$$2CsO_2 + 2H_2O \rightarrow 2CsOH + H_2O_2 + O_2$$

Cesium superoxide reacts with dry carbon dioxide only at elevated temperatures. The reaction proceeds as follows:

$$2CsO_2 + CO_2 \rightarrow Cs_2CO_3 + 1.5O_2$$

CsO_2 is reduced by hydrogen at 300°C with the formation of CsOH and water and liberation of oxygen. The reaction probably proceeds through the following stages:

$$2CsO_2 + 2H_2 \rightarrow Cs_2O_2 + 2H_2O$$

$$Cs_2O_2 + 2H_2O \rightarrow 2CsOH + H_2O_2$$

$$H_2O_2 \rightarrow H_2O + 0.5O_2$$

$$2CsO_2 + 2H_2 \rightarrow 2CsOH + H_2O + 0.5O_2$$

It has been noted [136] that CsO_2 will react with ozone to form cesium ozonide, CsO_3.

TABLE XI

Basic Properties of Alkali Metal Superoxides

Property	NaO(I)	α-KO_2	RbO_2	CsO_2
Active oxygen content, wt. %	43.6	33.8	20.4	14.5
Density, g/cm^3	2.21	2.158	3.06	3.80
Melting point, °C	—	440(β-KO_2)	412	450
Disintegration point, °C	> 540	660	> 567	> 597
ΔH°_{298}, kcal/mole	−62.1 ± 0.7	−67.6 ± 0.8	−63.0 ± 10.0	−62.0 ± 10.10
ΔF°_{298}, kcal/mole	−51.8	−57.4	−52.5 ± 11.0	−50.5 ± 11.5
ΔS°_{298}, eu	−16.8	−18.2	(−18)	(−18)
μeff, Bohr magn.	2.07	2.08	1.89	1.89
Lattice	fcc	tetragonal	tetragonal	tetragonal
U, kcal/mole	(190.9)	(173.2)	(165.5)	(158.8)

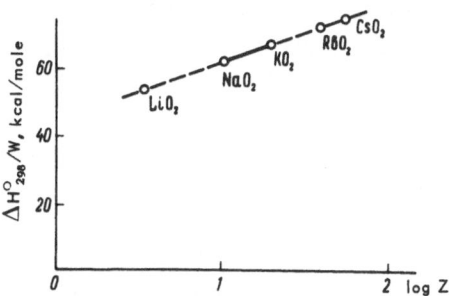

Fig. 19. ΔH°_{298} /W ratio for alkali metal
superoxides versus log Z.

The basic properties of the alkali metal superoxides are summarized in Table XI. The standard heats of formation of sodium superoxide and potassium superoxide have been determined with sufficient accuracy (see Table XI). By the application of Kapustinskii's logarithmic rule (see Fig. 19), values for the heats of formation of RbO_2 and CsO_2 can be estimated to be close to -72 and -74 kcal/mole. In the same manner, an estimated value of -53 kcal/mole is obtained for the heat of formation of LiO_2.

As can be seen from the data presented in Fig. 19, the heats of formation of the alkali metal superoxides increase as one goes from LiO_2 to CsO_2, which is in direct contrast to the case of the alkali metal peroxides (see Fig. 7). This difference is due to the fact that the lattice energy of an alkali metal superoxide is less than the lattice energy of the corresponding alkali metal peroxide. Moreover, the difference in the lattice energies of the superoxide is much smaller than the difference between the lattice energies of corresponding peroxides. For example, the difference in the lattice energies of NaO_2 and KO_2 is approximately 18 kcal/mole, which is about the same as the differences in the ionization potentials of Na and K (see Table IV), whereas the difference of the lattice energies of Na_2O_2 and K_2O_2 is approximately 50 kcal/mole (see Table III).

A comparison of the heats of formation of the oxides, peroxides, and superoxides of sodium and potassium [73], i.e.,

Na_2O — 99.4 ± 1.5 kcal/mole [73] K_2O — 86.4 ± 2 kcal/mole [73]
Na_2O_2—122.0 ± 1.2 kcal/mole K_2O_2—118.0 ± 10 kcal/mole
Na_2O_4—124.2 ± 1.4 kcal/mole K_2O_4—135.2 ± 1.6 kcal/mole

supports de Forcrand's empirical rule that the bonding of additional oxygen atoms to the alkali metal oxides reduces the thermal effects per single oxygen atom.

According to S. A. Shchukarev [137], the transition of oxides to superoxides takes place via the reaction

$$M_2(O^{2-}) + 3O \rightarrow 2M(O_2^-)$$

whereas the transition of peroxides to superoxides, according to Kazarnovskii [138], takes place via the reaction

$$M_2(O-O^{2-}) + O_2 \rightarrow 2M(O-O^-)$$

The Alkaline Earth Metal Superoxides – $Ca(O_2)_2$, $Sr(O_2)_2$, $Ba(O_2)_2$

Attempts to synthesize alkaline earth superoxides by the methods employed for the synthesis of alkali metal superoxides have not proved to be successful.

Combustion of metallic calcium, strontium, and barium in oxygen atmospheres results in the formation of the simple oxides of these metals. Attempts to prepare calcium, strontium, and barium superoxides, starting with the peroxides and using the same conditions under which NaO_2 is prepared from Na_2O_2, have also been unsuccessful [139]. Calculations performed by J. Margrave [35] indicate that $Ca(O_2)_2$ is thermodynamically unstable. Experiments performed by the author of this book have shown that the reactions of calcium, strontium, and barium peroxides with oxygen at superhigh pressures, 3300-3900 atm and 200-230°C, result in the formation of mixtures containing the superoxide. In the case of barium, the mixture contains approximately 10.4 wt.% $Ba(O_2)_2$ [140]. The fact that the yield of $Ba(O_2)_2$ could not be increased even when the reaction was run under 7000 atm O_2 pressure confirms Vannerberg's conclusion [125] that the formation of $Ba(O_2)_2$ results from the disproportionation of barium peroxide to the superoxide and oxide, rather than from a reaction involving the direct oxidation of the peroxide.

The author of this book has observed that the passage of oxygen or ozone through liquid ammonia solutions of metallic calcium and calcium salts causes the solutions to take on an intense red color which is indicative of the formation of higher oxides. However, the color disappears suddenly, and it has not been possible to separate higher oxides by evaporation of the liquid ammonia.

The reaction of KO_2 or NaO_2 with liquid ammonia solutions of $Ca(NO_3)_2$ and $Sr(NO_3)_2$ results in the formation of CaO_2 and SrO_2 [37]. It is felt that these peroxides are formed as a result of the decomposition of the unstable superoxides. The reaction of KO_2 with a liquid ammonia solution of $Ba(NO_3)_2$ yields a mixed oxide with the following composition [120]:

$$K_2^+ Ba^{2+} (O_2^-)_2 O_2^{2-}$$

It has been reported [141] and confirmed in subsequent studies [142-145] that the vacuum desiccation of the products obtained from the reaction of $CaO_2 \cdot 8H_2O$, $BaO_2 \cdot 8H_2O$, and $SrO_2 \cdot 8H_2O$ with hydrogen peroxide results in the formation of products containing up to 13% $M(O_2)_2$. Methods for the analysis of mixtures of CaO_2, $Ca(O_2)_2$, $Ca(OH)_2$, and $CaCO_3$ are described in references [142, 146, 147].

It has been proven that $M(O_2)_2$ is not formed directly from the reaction of the alkaline metal peroxide octahydrate with concentrated hydrogen peroxide. The superoxides form as a result of the disproportionation of $MO_2 \cdot 2H_2O_2$, which is formed in the $M(OH)_2-H_2O_2-H_2O$ system [143-145]. It has been established that the best yields of $Ca(O_2)_2$, $Sr(O_2)_2$, and $Ba(O_2)_2$ are obtained when disproportionation of the peroxide diperoxyhydrates proceeds at approximately 50°C and a residual pressure not exceeding 10 mm Hg. The reaction time is approximately 100 min. And, it is necessary to spread the sample in a thin layer over a very large surface. Under these

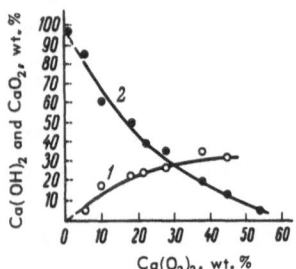

Fig. 20. Composition of the solid products from the disproportionation of $CaO_2 \cdot 2H_2O_2$. (1) Ca(OH) ; (2) CaO_2.

conditions, it is possible to obtain mixtures of MO_2, $M(OH)_2$, and $M(O_2)_2$ containing approximately 55 wt.% $Ca(O_2)_2$, 30 wt.% $Sr(O_2)_2$, and 11 wt.% $Ba(O_2)_2$ [148, 149]. It has been established that the disproportionation of peroxide diperoxyhydrates to superoxides may also be initiated with ultraviolet radiation [150], indicating that the process is directly related to the chain dissociation mechanism of the crystallized hydrogen peroxide [151]. The amounts of $M(O_2)_2$ obtained from these reactions are close to those theoretically expected, provided the reactions proceed as follows [151]:

$$2(MO_2 \cdot 2H_2O_2) \rightarrow M(O_2)_2 + M(OH)_2 + 3H_2O + 1.5O_2$$

Analysis of the products obtained from the vacuum desiccation of $CaO_2 \cdot 2H_2O_2$ indicates that as the amount of $Ca(O_2)_2$ increases, the amount of $Ca(OH)_2$ also increases, whereas the amount of CaO_2 decreases [106] (see Fig. 20). These facts can be explained either by the occurrence of a secondary reaction of $Ca(O_2)_2$ with water vapor formed from the decomposition of the H_2O_2, or, as has been suggested by Kazarnovskii [94], as a result of the participation of the hydroperoxyl radical in the reaction. It does not appear that it will be possible to obtain, from these reactions, a product containing more than 50 mol.% $M(O_2)_2$, since for each mole of $M(O_2)_2$ formed, one mole of $M(OH)_2$ is also formed.

Products containing up to 23 wt.% $Ca(O_2)_2$ were also obtained [125, 152] from the reaction of moist air at 200°C with CaO_2 or $CaO_2 \cdot 8H_2O$. It is unlikely that this process would take place without the hydrogen peroxide [151] formed via the hydrolysis of the calcium peroxide compounds. Formation of peroxide compounds containing up to 18% $Ca(O_2)_2$ results from the ozonization of $CaO_2 \cdot 2H_2O_2$.

Peroxide products containing calcium superoxide are paramagnetic [142-146]. They are stable when stored in hermetically sealed containers [149]. At approximately 290°C, they liberate the superoxide oxygen, and at 340°C, the peroxide oxygen is also liberated [140]. They rapidly react with water, liberating the superoxide oxygen [147, 153]. When dissolved in diluted acids, they react by liberating oxygen and hydrogen peroxide via the intermediate formation of the HO_2 radical [154]. These mixtures are also more effective CO_2 scrubbers than pure calcium peroxide [153]. In one of the recent U.S. patents [155], a method is described for the preparation of a mixture of $Ca(O_2)_2$ and an alkali metal halide via a reaction of the type

$$2KO_2 + CaCl_2 \rightarrow Ca(O_2)_2 + 2KCl$$

These mixtures are used for the passivation of iron and steel products.

The structure of the alkaline earth metal superoxide has not been determined since it has not been possible to separate them from products containing the peroxides. It is possible that they are only stable in a solid solution of peroxide [125]. According to Vannerberg [125], the mixtures may be similar to tungsten bronzes. There is some experimental support of this view based on the fact that x-ray powder photographs only show the peroxide phase [152].

The effective magnetic moment of $Ca(O_2)_2$ is 2.83 Bohr magnetons [146], which indicates the presence of two unpaired electrons. The lattice energy of $Ba(O_2)_2$ is estimated as 515.8 kcal/mole. The standard heat of formation, $\Delta H°_{298}$, is estimated as 166.1 kcal/mole, and $S°_{298}$ = 28.7 eu. For the reaction

$BaO_2 + O_2 \rightarrow Ba(O_2)_2$, $\Delta H^\circ_{298} = -14.2$ kcal, and $\Delta S^\circ_{298} = -45.4$ eu [140].

Superoxides of other elements are not known. The assumption that monovalent thallium superoxide, TlO_2, is formed as an anode deposit in the electrolysis of certain salts of thallium [156] has been discounted on the basis of more recent studies [157].

It has been reported that solid mixtures containing small amounts of $Mg(O_2)_2$, $Cd(O_2)_2$, and $Zn(O_2)_2$ are formed by the reaction of $Mg(OH)_2$, $Cd(OH)_2$, and $Zn(OH)_2$ with hydrogen peroxide [158–160].

In reference [161] a doubtful claim is made for the formation of trivalent gold superoxide, $Au(O_2)_3$, on a metallic gold surface heated to 500°C.

REFERENCES

1. R. Kirk and O. Othmer. Encyclopedia of Chemical Technology, Vol. 10, New York, Interscience Encyclopedia (1953), p. 38.
2. J. Clarke. J. Am. Soc. Naval Engrs. 68:105 (1956).
3. W. Schechter and R. Shakely. Handling and Use of the Alkali Metal, Advances in Chemistry, Ser. 19, ACS, Washington (1957), p. 124.
4. G. Minkoff. Frozen Free Radicals. Moscow, IL (1960).
5. D. J. Fabian. Seventh International Symposium on Combustion, London, Butterworths (1958), p. 150.
6. G. J. Minkoff and C. Tipper. Chemistry of Combustion Reactions, London, Butterworths (1962).
7. A. J. B. Robertson. Trans. Faraday Soc. 48:228 (1952).
8. S. N. Foner and R. L. Hudson. J. Chem. Phys. 21:1608 (1953).
9. K. Ingold. J. Chem. Phys. 24:360 (1956).
10. S. N. Foner and R. L. Hudson J. Chem. Phys. 21:1374 (1953).
11. F. P. Losing and A. W. Tickner. J. Chem. Phys. 20:907 (1952).
12. L. Kerwin and M. Cottin. Can. J. Phys. 36:184 (1958).
13. A. Robertson. Applied Mass Spectrometry, Gos. Nauch.-Tekhn. Izd. Neftyanoy i Gornotoplivnoy Literatury, Moscow (1958), p. 116.
14. S. N. Foner and R. L. Hudson. J. Chem. Phys. 23:1364 (1955).
15. S. N. Foner and R. L. Hudson. J. Chem. Phys. 36:2681 (1962).
16. S. N. Foner and R. L. Hudson. J. Chem. Phys. 23:1974 (1955).
17. R. Klein and M. Scheer. J. Chem. Phys. 31:278 (1959).
18. K. B. Harwey and N. Brown. J. Chem. Phys. 56:745 (1959).
19. S. Siegel and L. H. Baum. J. Chem. Phys. 32:1249 (1960).
20. J. Kroh, B. C. Green, and J. W. T. Spinks. J. Am. Chem. Soc. 83:2201 (1961).
21. I. I. Skorokhodov and V. B. Golubev. Zh. Fiz. Khim. 36:93 (1961).
22. E. Muschlitz and P. L. Bailey. J. Phys. Chem. 60:681 (1956).

23. W. Schumb et al. Hydrogen Peroxide, Moscow, IL (1958).
24. A.D. Walsh. J. Chem. Soc. (1953), p. 2288.
25. S.N. Foner. Free Radicals in Inorganic Chemistry, Advances in Chemistry, Ser. 36, ACS, Washington (1962), p. 43.
26. P. Gray. Trans. Faraday Soc. 55:408 (1959).
27. D.E. Milligan and M.E. Jacox. J. Chem. Phys. 38:2627 (1963).
28. Catalysis, Investigation of Homogenic Processes, Moscow, IL (1957), p. 96.
29. A.O. Allen. Radiation Chemistry of Water and Its Solutions, Moscow, Gosatomizdat (1963). [Published in English by D. Van Nostrand, Princeton, New Jersey.]
30. J. Jortner and G. Stein. Bull. Res. Council Israel A6:239-246 (1957).
31. E. Saito and B.H.J. Bielski. J. Am. Chem. Soc. 83:4467 (1961).
32. J. Kroh. Can J. Chem. 46:413 (1962).
33. P. Hartek. J. Chem. Phys. 22:1746 (1954).
34. J. Thompson and J. Kleinberg. J. Am. Chem. Soc. 73:1243 (1951).
35. Chem. Eng. News 31(39):4012 (1953).
36. A. Davidson and J. Kleinberg. J. Phys. Chem. 57:571 (1953).
37. D.L. Schechter and J. Kleinberg. J. Am. Chem. Soc. 76:3297 (1954).
38. I.I. Vol'nov and A.N. Shatunina. Izv. Akad. Nauk SSSR, Otd. Khim. Nauk (1957), p. 762.
39. I.I. Vol'nov and A.N. Shatunina. Zh. Neorgan. Khim. 2:257 (1959).
40. N.G. Vannerberg. Progress in Inorganic Chemistry, Vol. 4, New York, Interscience Publishers, Inc. (1962), p. 137.
41. I.A. Kazarnovskii. Izv. Akad. Nauk SSSR, Otd. Khim. Nauk (1949), p. 221.
42. E.B. Schoene. Experimental Investigation on Hydrogen Peroxide, Moscow, Tip Isleneva (1875), p. 131.
43. R. de Forcrand. Compt. Rend. 150:1399 (1910).
44. F. Haber and H. Sachse. Z. Phys. Chem. Bodenstein Festband (1931), p. 831.
45. I.I. Vol'nov and A.N. Shatunina. Zh. Neorgan. Khim. 4:1491 (1959).
46. G.L. Cunningham. U.S. Patent 2908552 (1959).
47. N.G. Vannerberg. Progress in Inorganic Chemistry, Vol. 4, New York, Interscience Publishers, Inc. (1962), p. 133.
48. W. Schechter, H. Sisler, J. Thompson, and J. Kleinberg. J. Am. Chem. Soc. 70:267 (1948).
49. W. Schechter, J. Thompson, and J. Kleinberg. J. Am. Chem. Soc. 71: 1816 (1949).
50. W. Stephanou, W. Schechter, and J. Kleinberg. J. Am. Chem. Soc. 71:1819 (1949).
51. J. Kleinberg. Unfamiliar Oxidation States and Their Stabilization, University of Kansas Press, Lawrence (1950), p. 25.
52. Inorganic Syntheses, Vol. 4, New York, McGraw-Hill (1953), p. 82.
53. S.H. Cohen, I.L. Margrave, V. Shelar, and E.M. Montaban. Inorg. Nucl. Chem. 14:301 (1960).
54. W.H. Schechter. U.S. Patent 2648596 (1953).
55. W.H. Schechter. Canadian Patent 505912 (1954).
56. I.A. Anderson and N.J. Clark. J. Phys. Chem. 67:2135 (1963).
57. J.F. Bedinger, N.S. Gosh, and E.R. Marning. Threshold of Space, Proc. Conf. Chem. Astronomy, 1956, Pergamon Press (1957).
58. R. Setton. French Patent 1179010 (1959).
59. A. Le Berre and P. Goasguen. Compt. Rend. 254:1306 (1962).
60. J.E. Bennett, D.Y. Ingram, and D. Schonland. Proc. Phys. Soc. 69:556 (1956).
61. E. Brame. J. Inorg. Nucl. Chem. 4:90 (1957).
62. T.V. Rode and G.A. Gol'der. Izv. Akad. Nauk SSSR, Otd. Khim. Nauk (1956), p. 299.

63. G.F. Carter and D.H. Templeton. J. Am. Chem. Soc. 75:5247 (1953).
64. D.H. Templeton and C.H. Dauben. J. Am. Chem. Soc. 72:2251 (1950).
65. G.S. Zhdanov and Z.V. Zvonkova. Dokl. Akad. Nauk SSSR 82:743 (1952).
66. S.S. Todd. J. Am. Chem. Soc. 75:1229 (1953).
67. A.B. Neiding and I.A. Kazarnovskii. Zh. Fiz. Khim. 24:1407 (1950).
68. T.V. Rode. Dokl. Akad. Nauk SSSR, 90:1077 (1953).
69. R.L. Tallmann. Dissertation Abstr. 20:4293 (1960).
70. K.B. Yatsimirskii. Izv. VUZOV, Khim. i Khim. Tekhnol. 2:480 (1959).
71. P.W. Gilles and J.L. Margrave. J. Phys. Chem. 60:1333 (1956).
72. L. Brewer. Chem. Rev. 52:6 (1953).
73. J. Coughlin. Bull. 542 Bureau of Mines, Washington (1954), p. 46.
74. J.L. Margrave. J. Chem. Educ. 32:522 (1955).
75. S.H. Cohen and J.L. Margrave. Anal. Chem. 29:1462 (1957).
76. A. Kh. Mel'nikov and T.P. Firsova. Zh. Neorgan. Khim. 4:169 (1961).
77. E. Seyb and J. Kleinberg. Anal. Chem. 23:115 (1951).
78. A. Kh. Mel'nikov and T.P. Firsova. Zh. Neorgan. Khim. 6:2230 (1961).
78a. A. Meffert and H. Meier–Ewert. Z. Anal. Chem. 198:325 (1963).
79. A.B. Tsentsiper and S.A. Tokareva. Zh. Neorgan. Khim. 6:2474 (1961).
80. T.V. Rode and A.V. Zachatskaya. Zh. Neorgan. Khim. 5:524 (1960).
81. T.V. Rode and G.K. Grishenkova. Zh. Neorgan. Khim. 5:529 (1960).
82. T.V. Rode and G.A. Gol'der. Zh. Neorgan. Khim. 5:535 (1960).
83. Hydrogen Peroxide and Peroxide Compounds, published by M.E. Pozin, Moscow, Goskhimizdat (1951).
84. H. Lux. Z. Anorg. Allgem. Chem. 298:285 (1959).
85. F. Onuska. Chem. Zvesti 14:459 (1960).
86. C.B. Jackson. U.S. Patent 2405580 (1946).
87. Mine Safety Appliance Co. British Patent 626644 (1946).
88. R.R. Miller, U.S. Patent 2414116 (1947).
88a. R.M. Bovard. Aerospace Med. 31:407 (1960).
89. Mine Safety Appliance Co. British Patent 629406 (1949).
90. H. Ostertag and E. Rink. Compt. Rend. 234:958 (1952).
91. M.I. Klyashtornyi. Zh. Prikl. Khim. 32:337 (1959).
92. R. Kohlmu."er. Ann. Chim. 4:1202 (1959).
93. L. Linemann and G. Tridot. Compt. Rend. 236:1282 (1953).
94. I.A. Kazarnovskii and A.B. Neiding. Dokl. Akad. Nauk SSSR 86:717 (1952).
95. A.W. Petrocelli and D.L. Kraus. J. Chem. Educ. 40:146 (1963).
96. A. Le Berre. Compt. Rend. 252:1341 (1961).
97. A. Le Berre. Bull. Soc. Chim. France (1961), p. 1543.
98. A. Le Berre. Bull. Soc. Chim. France (1962), p. 1682.
99. C. Kroger. Z. Anorg. Allgem. Chem. 253:92 (1945).
100. F. Fischer and H. Ploetze. Z. Anorg. Allgem. Chem. 75:30 (1912).
101. A. Kh. Mel'nikov and T.P. Firsova. Zh. Neorgan. Khim. 8:560 (1963).
102. J. Weiss. Trans. Faraday Soc. 31:673 (1935).
103. P. Selwood. Magnetochemistry, Moscow, IL (1958), p. 279. [Second rev. ed. published in English by Interscience Publishers, Inc., New York.]
104. T.V. Rode. Thesis Reports of the Third Conference on Physico–Chemical Analysis, Moscow, Izd. Akad. Nauk SSSR (1955), p. 110.
105. M. Blumenthal. Roczniki Chem. 12:127 (1932).
105a. R. de Forcrand. Compt. Rend. 158:991 (1914).
106. E.I. Sokovnin. Izv. Akad. Nauk SSSR, Otd. Khim. Nauk (1963), p. 181.
107. V. Kasatochkin and V. Kotov. Zh. Tekhn. Fiz. 7:1468 (1937).
108. A. Helms and W. Klemm. Z. Anorg. Allgem. Chem. 241:97 (1939).
109. S.C. Abrahams and J. Kalnais. Acta Cryst. 8:503 (1955).
110. I.A. Kazarnovskii and S.I. Raikhshtein. Zh. Fiz. Khim. 21:245 (1947).

111. F. Halverson. Phys. Chem. Solids 23:207 (1962).
112. C. F. Carter and J. L. Margrave. Acta Cryst. 5:851 (1952).
112a. Science in World War II, OSRD Chemistry, edited by W.A. Noyes, Jr., Boston (1948), p. 363.
113. A. Kh. Mel'nikov and T.P. Firsova. Zh. Neorgan. Khim. 7:1228 (1962).
114. E. Wyss-Dunant. Bull. Schweiz. Akad. Med. Wiss. 9:221 (1953).
115. C.B. Jackson, H.C. Beam, and A. Van Andel. U.S. Patent 2494131 (1950).
116. C.B. Jackson and A. Van Andel. U.S. Patent 2517209 (1950).
117. R.M. Bovard. U.S. Patent 2889210 (1959).
118. R.M. Bovard. U.S. Patent 2913317 (1959).
118a. R.M. Bovard. U.S. Patent 2758015 (1956).
118b. D.K. Dieterly et al. C.A. 59:13577 (1963).
119. I.I. Vol'nov, E.I. Sokovnin, V.V. Matveev. Izv. Akad. Nauk SSSR, Otd. Khim. Nauk (1962), p. 1127.
120. E. Seyb and J. Kleinberg. J. Am. Chem. Soc. 73:2308 (1961).
121. M. Schmidt and H. Bipp. Z. Anorg. Allgem. Chem. 303:190 (1960).
122. M. Schmidt and H. Bipp. Z. Anorg. Allgem. Chem. 303:201 (1960).
123. P.A. Giguère and K.B. Harvey. J. Am. Chem. Soc. 76:5891 (1954).
124. D.I. Mendeleev. Principles of Chemistry, Pt. 2, 2nd Edition, St. Petersburg, 1871, Works, Vol. 14, Moscow, Izd. Akad. Nauk SSR (1949), p. 78.
125. N.G. Vannerberg. Progress in Inorganic Chemistry, Vol. 4. New York, Interscience Publishers, Inc. (1962), p. 128.
126. A. Helms and W. Klemm. Z. Anorg. Allgem. Chem. 242:201 (1939).
127. D.L. Kraus and A.W. Petrocelli. Dissertation Abstr. 21:1081 (1960).
128. F. Kuhbier. Beiträge zur Kenntnis der Peroxyde und Tetraoxyde des Rubidiums, Caesiums und Tetrametilammoniums, Dissertation, Berlin(1929).
129. E. Rengade. Bull. Soc. Chim. Paris 35(3):769 (1906).
130. E. Rengade. Compt. Rend., 142:1150 (1906).
131. E. Rengade. Ann. Chim. Phys. 11(6):348 (1907).
132. I.I. Vol'nov and V.V. Matveev. Zh. Prikl. Khim. (in press).
133. P.G. Borisyak. Zh. Tekhn. Fiz. 20:928 (1950).
134. M. Centnerszwer and M. Blumenthal. Bull. Acad. Pol. 1933(A):498.
134a. G.V. Morris. Dissertation Abstr. 23:2343 (1963).
135. L.I. Kazarnovskaya. Zh. Fiz. Khim. 20:1403 (1946).
136. I.I. Vol'nov and V.V. Matveev. Izv. Akad. Nauk SSSR, Otd. Khim. Nauk (1963), p. 1136.
137. S.A. Shchukarev. Zh. Obshch. Khim. 28:857 (1958).
138. I.A. Kazarnovskii. Zh. Fiz. Khim. 14:333 (1940).
139. E. Seyb and J. Kleinberg. J. Am. Chem. Soc. 73:2308 (1951).
140. I.I. Vol'nov and A.N. Shatunina. Dokl. Akad. Nauk SSSR 110:87 (1956).
141. D. Baumann. Iowa State Coll. J. Sci. 28:280 (1954).
142. I.I. Vol'nov, V.N. Chamova, and V.P. Sergeeva. Zh. Neorgan. Khim. 1:1937 (1956).
143. I.I. Vol'nov and V.N. Chamova. Zh. Neorgan. Khim. 2:263 (1957).
144. I.I. Vol'nov and E.I. Latysheva. Zh. Neorgan. Khim. 2:259 (1957).
145. I.I. Vol'nov and E.I. Latysheva. Zh. Neorgan. Khim. 2:1697 (1957).
146. R.S. Johnston, E.D. Osgood, and R.R. Miller. Anal. Chem. 30:511 (1958).
147. I.I. Vol'nov and E.I. Latysheva. Zh. Analit. Khim. 14:242 (1959).
148. I.I. Vol'nov and V.N. Chamova. Zh. Neorgan. Khim. 3:1098 (1958).
149. I.I. Vol'nov and V.N. Chamova. Zh. Neorgan. Khim. 4:253 (1959).
150. I.I. Vol'nov and V.N. Chamova. Zh. Neorgan. Khim. 3:1095 (1958).
151. I.I. Vol'nov and V.N. Chamova. Zh. Neorgan. Khim. 5:522 (1960).
152. C. Brosset and N.G. Vannerberg. Nature 177:238 (1956).
153. I.I. Vol'nov and A.N. Shatunina. Zh. Neorgan. Khim. 2:1474 (1957).

154. K. V. Astakhov and A. G. Getsov. Dokl. Akad. Nauk SSSR 81:43 (1951).
155. J. S. Hashmann. U.S. Patent 3119665 (1964).
156. M. Centnerszwer and T. Trebaczkiewitz. Z. Phys. Chem. 165:367 (1933).
157. M. S. Skanavi-Grigor'eva and A. N. Staroverova. Zh. Obshch. Khim. 28:1689 (1958).
158. N. G. Vannerberg. Arkiv Kemi 14:100 (1959).
159. N. G. Vannerberg. Arkiv Kemi 10:455 (1957).
160. N. G. Vannerberg. Arkiv Kemi 14:119 (1959).
161. N. A. Shishakov. Zh. Fiz. Khim. 31:37 (1957).

Chapter Five

Alkali Metal Ozonides

In the course of studies on the reaction of ozone with anhydrous metal hydroxides, I. A. Kazarnovskii [1, 2] confirmed the formation of red-colored products which had been first reported at the turn of the century by a number of investigators [3, 4, 5]. These products represent a new class of compounds characterized by the presence of the O_3^- ion and classified as ozonides. The O_3^- ion has an odd number of electrons and, as a result, the ozonides are paramagnetic. They can be considered to be free radicals with high chemical reactivity and an unusally long lifetime. The existence of the O_3^- ion in these compounds has been established by x-ray studies [6], magnetic studies [1], electron paramagnetic resonance studies [7-10], and by spectroscopic studies [7].

The ozonides of sodium, potassium, rubidium, cesium, and ammonium have been prepared. Attempts to synthesize pure ozonides of lithium and the alkaline earth metals have not, as yet, been successful.

According to Kazarnovskii [2, 11], the formation of alkali metal ozonides proceeds according to the following steps:

$$2MOH + O_3 \rightarrow 2MOH \cdot O_3 \tag{a}$$

$$2MOH \cdot O_3 + 2O_3 \rightarrow 2MO_3 + 2HO_2 + 2O_2 \tag{b}$$

$$2HO_2 \rightarrow H_2O + 1.5O_2 \tag{c}$$

$$MOH + H_2O \rightarrow MOH \cdot H_2O \tag{d}$$

$$3MOH + 4O_3 \rightarrow 2MO_3 + MOH \cdot H_2O + 3.5O_2 \tag{e}$$

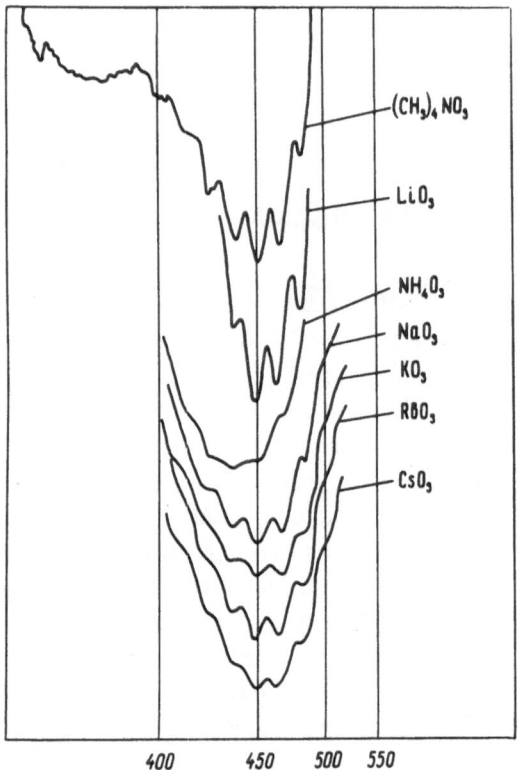

Fig. 21. Visible spectra of alkali metal ozonides
dissolved in liquid ammonia [20].

Consequently, a mixture is formed of the ozonide and the alkali metal hydroxide hydrate with the simultaneous transformation of a large amount of ozone to oxygen. The activation energy of the KOH reaction with ozone is estimated as 3.1 kcal [12].

The alkali metal ozonides are soluble in such solvents as liquid ammonia, dimethylformamide, and methylamine. The solubility of KOH and NaOH in these solvents is very small. For example, in liquid ammonia at –40°C, KOH is soluble only to $2.8 \cdot 10^{-5}$ wt.% [13], and NaOH only to $2.5 \cdot 10^{-4}$ wt.% [14]. It is thus possible to extract essentially pure KO_3 and NaO_3 from the reaction product. Liquid ammonia extractions have yielded 98% pure KO_3 and 91% pure NaO_3 [1, 2, 15, 16]. Since the solubility of RbOH and CsOH is approximately 1 g per 100 ml of liquid ammonia at -40°C, the RbO_3 and CsO_3 obtained by means of liquid NH_3 solvent extraction are only approximately 67% pure [2].

The ozonization of sodium hydroxide at -60°C also results in the formation of NaO_3 [2, 17]. The hydroxides of potassium, rubidium, and cesium have been ozonized at temperatures from -60°C to room temperature [17-19].

As seen in Fig. 21, the spectra of liquid ammonia solutions of the five alkali metal ozonides and tetramethylammonium ozonide each show a maximum at 450 mμ [20], regardless of the cation type.

Lithium Ozonide – LiO₃

Pure lithium ozonide has not been synthesized. However, the tetraammoniate of lithium ozonide, $Li(NH_3)_4O_3$, has been prepared by the reaction of dilute ozone (3% ozone in oxygen) with anhydrous powdered lithium hydroxide containing ammonia adsorbed on its surface [20]. The ratio of LiOH to NH_3 was 5 g:0.2 g and the reaction was carried out at -112°C. The ozonide was extracted from the hydroxide with liquid ammonia, and it was salted out from the liquid ammonia with fluoroform. The composition of the deposited red solid product was determined as $Li(NH_3)_4O_3$ mixed with lithium nitrate and lithium nitrite. The standard heat of formation of lithium ozonide tetraammoniate is -135 ± 5 kcal/mole. It is possible that the main reason for the instability of lithium ozonide is the small radius of the lithium ion, i.e., 0.60 A. Since the complex ion $[Li(NH_3)_4]^+$ is larger than the Li^+ ion, it can form a compound with the O_3^- ion.

Sodium Ozonide – NaO₃

In the very few studies published prior to 1951 [3-5], it was noted that in the reaction of sodium hydroxide and ozone–oxygen mixtures, the hydroxide acquired a yellow color which suddenly faded away at room temperature. Kazarnovskii et al. [2], investigating the reaction of anhydrous sodium hydroxide powder and ozone—oxygen mixtures at -50 to -60°C, observed that, indeed, the hydroxide acquired an intense yellow color which, when extracted with liquid ammonia, resulted in the formation of a dark red solution. Upon evaporation of the ammonia, crystals were recovered. Analysis of the recovered product showed 90 wt.% NaO_3, 2-3% NaOH, and 4-6% water, as water of hydration. Kazarnovskii [2] further reported that NaO_3 is unstable, and that at room temperature it decomposes in 53 hr via the reaction

$$NaO_3 \rightarrow NaO_2 + \tfrac{1}{2}O_2$$

Appearing simultaneously with the reported NaO_3 work of Kazarnovskii was the report of work carried out in the United States by Whaley and Kleinberg [15]. Their report generally confirms the findings reported by Kazarnovskii. They noted, however, that the ozonization of NaOH can be caused to take place at room temperature, as well as at low temperatures. The NaO_3 obtained from the room temperature reaction is not soluble in liquid ammonia, and the ozonide—hydroxide mixture is quite stable. The presence of NaO_3 in the mixture was established by means of magnetic measurements. Subsequent studies [7, 16, 21, 22] have confirmed the room temperature synthesis of NaO_3. In order to explain the difference in the solubility and stability characteristics of NaO_3 prepared at room temperature and low temperatures, the idea has been proposed that NaO_3 can exist in two crystalline forms, one being soluble in liquid ammonia but unstable at room temperature, and the other being insoluble and stable [16]. As yet, it has not been positively established that two such crystal forms exist.

The data published on the structure of NaO_3 have been reviewed [23]. The NaO_3 lattice is tetragonal, with a = 11.61 A and c = 7.66 A [22]. The space group is 14/mmm [23a]. Its density varies from 1.56 to 1.60 g/cm^3 [23b]. The standard heat of formation of NaO_3 is estimated as 45 kcal/mole [2]. The differential heating curve for sodium ozonide shows an exothermal effect at -10°C, corresponding to its decomposition via the reaction [24]

$$NaO_3 \rightarrow NaO_2 + \tfrac{1}{2}O_2$$

Potassium Ozonide – KO₃

Production Methods. The method used for the synthesis of pure potassium ozonide is the same as that used for the synthesis of sodium ozonide [1]. Attempts have been made to improve the original method of synthesis and to find compounds, other than the hydroxides, capable of forming potassium ozonide.

According to references [18, 19], low temperature is not a necessary condition for the preparation of potassium ozonide via the ozonization of the hydroxide. The ozonide is also formed at +5°C [19] and even at +40°C [18]. Potassium ozonide can also be produced by ozonization of a potassium alcoholate [25, 26] and of potassium superoxide [26, 27]. Ozonization of potassium t-butoxide and t-perbutoxide results in the formation of KO_3 with yields of approximately 10 and 40%, respectively. However, it is not possible to separate the ozonide from the reaction mixture with liquid ammonia since the alcoholates are also soluble in liquid ammonia [25, 26]. Investigations conducted in the Laboratory of Peroxide Compounds (of the Institute of General and Inorganic Chemistry) [27] have shown that the ozonization of KO_2 at 40°C results in the formation of KO_3 with the same yield as is achieved from the ozonization of the hydroxide. The reaction proceeds as follows [2]:

$$KO_2 + O_3 \rightarrow KO_3 + O_2 + 28.5 \text{ kcal}$$

Potassium ozonide is also formed by the ozonization of 40% KOH solution at -40°C [7], by the passage of fluorine through concentrated KOH solutions at -20°C [28], and by the electrolysis of KOH–freon suspensions [30].

Properties of Potassium Ozonide. Potassium ozonide is a red, crystalline substance, stable only when stored in hermetically sealed containers at low temperatures. On standing it decomposes to the superoxide via the reaction [31]

$$2KO_3 \rightarrow 2KO_2 + O_2 + 11.0 \text{ kcal}$$

The kinetics of this reaction have been studied. The reaction is autocatalytic and characterized by an "induction period" which is 1.67 days at 18°C, 20 days at 0°C, 54 days at -9°C, and 205 days at -18°C [31]. In the ensuing active period the decomposition rate abruptly increases. At 50 to 60°C, the decomposition is complete in 15 min [2]. Electron paramagnetic resonance studies of the decomposition of potassium ozonide at 22°C have been made which indicate that the decomposition follows an exponential path with a time constant of 0.02 hr [18]. From differential thermal analysis, it is found that above $60 \pm 2°C$ KO_3 completely transforms to KO_2 [32]. Studies of the decomposition of KO_3 at 18 to 20°C and at 0°C by means of magnetic measurements indicate that the reaction proceeds with the intermediate formation of atomic oxygen as follows [33]:

$$KO_3 \rightarrow KO_2 + O$$

$$KO_3 + O \rightarrow KO_2 + O_2$$

$$\overline{}$$

$$2KO_3 \rightarrow 2KO_2 + O_2$$

Potassium ozonide is paramagnetic [33, 34]. Its magnetic susceptibility, $\chi_g \cdot 10^6$, is 16.9 ± 0.15, and the magnetic moment is 1.73 Bohr magnetons. Similarities between the x-ray powder patterns of potassium azide and potassium ozonide led to the conclusion that the crystal lattice of KO_3 is body-centered tetragonal of the potassium bifluoride type, similar to KN_3,

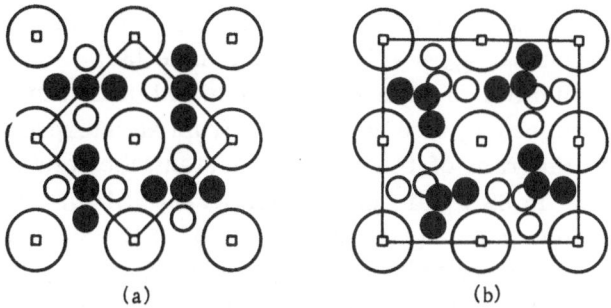

(a)   (b)

Fig. 22. (a) Potassium azide lattice; (b) potassium ozonide lattice [34].

and that the O_3^- ion accordingly is linear [6]. However, subsequent investigators questioned this interpretation [32, 35, 36], and finally in 1963 a detailed x-ray study of KO_3 was reported which established that the KO_3 lattice is body-centered tetragonal of the cesium chloride type with a = 8.597 ± 0.002 A and c = 7.080   0.002 A. The space group is 14/mcm [37]. The structures of KN_3 and KO_3 are compared in Fig. 22. It can be seen that the N_3^- ions, represented by the open and closed circles, are linear, whereas the O_3^- ions are bent or angular [37] ($\angle O-O-O = 100°$). As a result, each potassium ion in the KO_3 lattice is not equally coordinated by eight oxygen atoms and one must consider two kinds of potassium ions. The interatomic distance is 3.03 A for K_I -O_{II} and 2.70 A for K_{II} -O_{II} [37].

The density of KO_3 is 1.99 g/cm^3 [23b]. It is of interest to compare the densities of the peroxides, superoxides, and ozonides of sodium and potassium and to note that in both cases the ozonides have the lowest density:

Na_2O_2, 2.60 g/cm^3	K_2O_2, 2.40 g/cm^3
NaO_2, 2.21 g/cm^3	KO_2, 2.158 g/cm^3
NaO_3, 1.6 g/cm^3	KO_3, 1.99 g/cm^3

The standard heat of formation of KO_3 is 62.1 ± 0.9 kcal/mole and $\Delta F°_{298} = -45.5 \pm 3$ kcal/mole [38].

The solubility of potassium ozonide in liquid ammonia has been determined by two methods. From an interpretation of the phase diagram of the $NH_3–KO_3$ system, the solubility was found to be 14.82 g per 100 g of NH_3 at -35°C [39]. Measurements of the vapor pressure of liquid $NH_3–KO_3$ solutions indicate that the solubility of KO_3 at -63.5°C is 12 g per 100 g of NH_3. Potassium ozonide is a strong oxidizer, and 46% of its weight is active oxygen. It reacts spontaneously with water at room temperature and at 0°C with the liberation of oxygen and the formation of KOH. The reaction is believed to proceed as follows:

$$2KO_3 + 2H_2O \rightarrow 2KOH + 2OH + 2O_2$$

$$2OH \rightarrow H_2O + 0.5O_2$$

$$\overline{}$$

$$2KO_3 + H_2O \rightarrow 2KOH + 2.5O_2$$

Potassium ozonide reacts with acids with the formation of the OH radical and the liberation of O_2 [40], i.e.,

$$KO_3 + HX \rightarrow KX + OH + O_2$$

Studies of the isotopic exchange of oxygen between the OH free radical and water have been carried out by Kazarnovskii [41]. Reactions of KO_3 with formic acid and acetic acid have been studied. With aqueous solutions of formic acid, the products of the reaction include carbon dioxide, oxygen, and hydrogen peroxide. With anhydrous formic acid, the reaction products include carbon dioxide, oxygen, carbon monoxide, and hydrogen [42]. At 50°C, KO_3 reacts spontaneously with acetic acid vapor, and the reaction is characterized by a blue glow. The main products of this reaction are CO_2, CO, H_2, H_2O, and a small amount of methane [43]. Potassium ozonide does not react with F_2O_3 dissolved in freon-13 in the temperature range of -183 to -103°C. F_2O_3 decomposes to fluorine and oxygen [43a].

Rubidium and Cesium Ozonides — RbO_3, CsO_3

Pure cesium ozonide, CsO_3, has been obtained via the reaction of cesium superoxide and an ozone–oxygen gas mixture [18, 41]. Liquid ammonia extraction yielded a product containing 94.5 wt.% CsO_3 [44]. Thermal studies have revealed the onset of an exothermal process at 70°C. The process achieves its maximum rate at 100°C and corresponds to the decomposition of CsO_3 to CsO_2. Figure 23 shows the differential heating curves for the ozonides of sodium, potassium, and cesium. From these curves it can be seen that the thermal stability of the ozonides increases as the atomic number of the alkali metal increases.

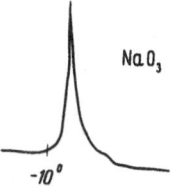

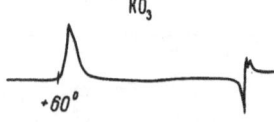

Fig. 23. Differential heating curves of sodium, potassium, and cesium ozonides.

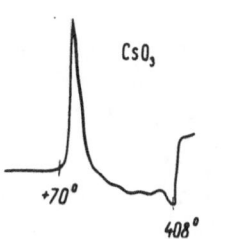

Ammonium and Tetramethylammonium Ozonides — NH_4O_3, $(CH_3)_4NO_3$

Ammonium ozonide, NH_4O_3, was first synthesized in 1962 [34]. It was obtained by the ozonization of liquid ammonia at temperatures below -100°C. Ammonium ozonide is stable below -126°C. At higher temperatures, it decomposes as follows:

$$2NH_4O_3 \rightarrow NH_4NO_3 + \tfrac{1}{2}O_2 + 2H_2O$$

Tetramethylammonium ozonide, $(CH_3)_4NO_3$, has also been synthesized. This compound was obtained [17] by the ozonization of tetramethylammonium hydroxide at + 25°C. Its solubility in liquid ammonia at -63°C is 1.3 ± 0.1 g per 100 g of NH_3. The compound is stable up to 25°C. The standard heat of formation of tetramethylammonium ozonide is 49.5 ± 4.5 kcal/mole.

Table XII summarizes some of the basic properties of the alkali metal ozonides.

TABLE XII

Basic Properties of Alkali Metal Ozonides

Formula	Active oxygen content, wt. %	Density, g/cm^3	Thermal stability limit, °C	$\Delta H°_{298}$, kcal/mole	Lattice	g Factor
(LiO_3)	72.7	—	—	—	—	—
NaO_3	56.3	1.6	−10 ± 2	—	Tetragonal body content	
KO_3	46.0	1.99	+60 ± 2	−62.1 ± 0.9	Same	2.0124
RbO_3	30.0	—	—	—	—	—
CsO_3	22.1	—	+70 ± 2	—	—	—
NH_4O_3	12.1	—	−126	—	—	2.0119

The Reaction Mechanism for the Formation of Alkali Metal
Ozonides

The reaction mechanism for the formation of the alkali
metal ozonides that is given on the first page of this chapter
indicates that 3 moles of hydroxide and 4 moles of ozone should
react to produce a 2:1 mixture of the ozonide and the hydroxide
monohydrate [2, 11]. This mechanism has not been proven
experimentally. It is based on two hypotheses: one concerns
the formation of an unstable complex, $MOH \cdot O_3$, (reaction a)
and the other concerns the role of the HO_2 radical (reaction b).
According to reaction (b), the hydroxide oxygen participates in
the formation of the HO_2 radical, i.e.,

$$M | \overline{OH \cdot O} | OO \rightarrow HO_2 + O_2$$

According to Foner [45], the only experimentally proven
reactions are reaction (c) and reaction (d). However, it
should be noted that at the ozonization temperatures (-15°C for
KOH, and -60°C for NaOH), and on the basis of the amount of
water generated in the system via reaction (c), it is impossible
that all of the KOH in the reaction product is in the form of the
monohydrate [46]. $KOH \cdot H_2O$ is stable at temperatures above
33°C, whereas $NaOH \cdot H_2O$ is stable only at 12°C [46]. Consequently, reaction (e) does not correspond to experimental
findings. It should be written in the following form:

$$nMOH + mO_3 \rightarrow (n - 3) MOH + MOH \cdot H_2O$$

$$+ 2MO_3 + (1.5m - 2.5)O_2$$

From reaction (e), assuming the formation of the hydroxide
monohydrate, the molar ratio of ozonide to hydroxide in the
final product should be 2:1. This means that for sodium
hydroxide, the ozonide content should be 71 wt.%, and for
potassium hydroxide the ozonide content should be 66.4 wt.%.
The claim made in reference [19] for the synthesis of an
hydroxide–ozonide mixture containing 63.5 wt.% KO_3 is doubtful.

In experiments with sodium hydroxide, it has been found that the ozonide content does not exceed 3-4 wt.% NaO_3 [22, 24], and with potassium hydroxide it does not exceed 4-5 wt.% KO_3 [18]. These results indicate that the ozonization of alkali metal hydroxides takes place mainly on the surface of the solid hydroxides. Consequently, the initial elementary processes in the mechanism for the formation of ozonides must be related to the catalytic decomposition of ozone.

The decomposition of ozone on the surface of KOH is very intense, especially at elevated temperatures [12]. Based on studies of the catalytic decomposition of ozone on different oxide surfaces, it appears that the process proceeds as follows:

$$O_3 + \text{catalyst} \rightarrow O_2 + O \cdot \text{catalyst} \tag{f}$$

The atomic oxygen is attached to the catalyst surface. The number of oxygen atoms formed as a result of the decomposition of ozone should be equal to the number of molecules of ozone decomposed. However, in the decomposition of a given number of ozone molecules, the catalyst does not necessarily attach all the oxygen atoms formed [12]. The number of oxygen molecules formed as a result of the decomposition of ozone should also be equal to the number of ozone molecules that are decomposed. However, this is not the situation according to reaction (e), in which the decomposition of 4 ozone molecules results in the formation of 3.5 oxygen molecules and 6 oxygen atoms (bonded in the form of $2MO_3$).

In an effort to determine if the mechanism formulated by Kazarnovskii [2, 11] on the basis of his studies at low temperatures is applicable to the formation of KO_3 at higher temperatures, the author of this book carried out isotopic studies at +40°C.

If reactions (a), (b), and (c) are correct, then it is to be expected that ozonide resulting from the ozonization of $KO^{18}H$ should not contain the heavy oxygen isotope since it was reported that the decomposition of the MOH $\cdot$ O_3 complex (reaction b) proceeds as follows [2, 11]:

$$M \overline{|O^{18}H \cdot O^{16}|} O^{16}O^{16} + O_3^{16} \rightarrow MO_3^{16} + HO^{18}O^{16} + O_2^{16}$$

Thus, the tagged atoms are to be found only in the water and in the oxygen that is formed as a result of the recombination of the HO_2 radicals, i.e.,

$$2HO_2 \rightarrow H_2O^* + 1.5O_2^*$$

However, as a result of the aforementioned studies carried out by this author at +40°C, it has been established that the oxygen of the $KO^{18}H$ molecules does indeed participate in the formation of potassium ozonide. The results of these studies are summarized in Table XIII. It is concluded, therefore, that if reactions (a) and (b) do play a role in the mechanism for the formation of ozonides, they do so only at low temperatures [2, 11].

In 1962 [27], the author of this book postulated the reaction

$$KO_2 + O_3 \rightarrow KO_3 + O_2 \tag{g}$$

TABLE XIII

Isotope Oxygen Balance in the Ozonization Reaction of $KO^{18}H$ with Ozone–Oxygen Mixture with Natural Isotope Composition (O^{18} in atomic %)

Initial KOH	KOH after ozonide extraction	Oxygen of potassium ozonide
	Expected content in reaction (b) by author [11]	
	1.370	In "ozonide" oxygen 0.204
		In "superoxide" and "peroxide" oxygen 0.204
1.370		In "oxide" oxygen 0.204
	Experimental findings	
	0.850	In "ozonide" oxygen 0.400
		In "superoxide" and "peroxide" oxygen 0.332
		In "oxide" oxygen 0.366

to account for the formation of potassium ozonide from the ozonization of potassium superoxide. Based on the occurrence of reaction (g) and the information developed by foreign scientists concerning the catalytic decomposition of ozone on a KOH surface [12, 26], it is possible to arrive at a mechanism for the formation of KO_3 in which atomic oxygen plays a vital role. Such a mechanism has been formulated by this author to be as follows:

$$3O_3 + 2KOH \rightarrow 3O_2 + 2KOH \cdot O + O \qquad \text{(h)}$$

$$2KOH \cdot O + O \rightarrow 2KO_2 + H_2O \qquad \text{(i)}$$

$$2O_3 + 2KO_2 \rightarrow 2KO_2 \cdot O + 2O_2 \qquad \text{(j)}$$

$$2KO_2 \cdot O \rightarrow 2KO_3 \qquad \text{(k)}$$

$$\overline{}$$

$$5O_3 + 2KOH \rightarrow 5O_2 + 2KO_3 \quad H_2O \qquad \text{(l)}$$

Reaction (h) is supported by the data given in reference [12]. Reactions (j) and (k) are supported by reaction (g), and experimental evidence for the occurrence of reaction (i) was obtained by us in the course of studies on the reaction of KOH with atomic oxygen generated in a glow discharge reaction. Thus, the overall formation reaction (l) is based on three experimentally proven intermediate reactions and corresponds to what is known concerning the decomposition of ozone on a KOH surface, i.e., the number of oxygen molecules formed is the same as the number of oxygen atoms formed. The water formed in the reaction is bonded to the unreacted KOH.

Alkaline Earth Metal Ozonides — $M(O_3)_2$

Very little detailed experimental work has been done on the synthesis of the alkaline earth metal ozonides, and the question concerning formation of such ozonides remains open.

E. Manchotte and B. Kampschulte [47] observed that in the course of the room temperature reaction of the oxides of magnesium, calcium, strontium, and barium with an ozone–oxygen mixture containing 8% ozone, the ozone very rapidly decomposes.   At −80°C, barium oxide acquires an intense yellow color.   With calcium oxide the color is less intense, and with magnesium oxide there is hardly any color developed at all.   Manchotte assumed these colored compounds to be ozonides.   However, this assumption has not been confirmed.   M. Blumenthal [48] indicates that the reaction of magnesium and strontium oxides with ozonized oxygen produces, at room temperature and at −80°C, a substance which does not react with potassium permanganate solution, but does liberate iodine from an acidic KI solution.   This evidence was given as support for the formation of ozonides; however, the possibility that the liberation of iodine results from the presence of ozone absorbed on the oxides' surfaces has not been discounted.   G. Schwab and G. Hartman [12] studied the catalytic decomposition of ozone on the surfaces of magnesium and barium oxides and barium peroxide at 20-150°C without detecting any changes in the chemical composition of these compounds.   Experiments by the author of this book on the ozonization of calcium, strontium, and barium oxides at −70°C, as well as the hydroxides and peroxides of these elements in the temperature range of +20 to −70°C, did not reveal any noticeable formation of compounds differing from the initial substances.

Attempts were made to synthesize alkaline earth metal ozonides by oxidizing, with oxygen or ozone, liquid NH_3 solutions of metallic calcium, strontium, and barium.   A. Guntz and B. Mentrel [49] dissolved barium in liquid ammonia at -35 to -50°C and oxidized the blue solution.   They obtained a product containing BaO and 7.5 to 9.0 wt.%. BaO_2.   J. Thompson and J. Kleinberg [50] observed that the ozonization of the blue liquid NH_3–calcium solution results in the formation of a gray product, the analysis of which was not reported. W. Strecker and M. Thienemann [5] ozonated liquid ammonia solutions of barium and calcium. With the calcium solution they obtained a brown deposit which completely disintegrated after the ammonia was evaporated. Experiments conducted in the Laboratory of Peroxide Compounds (of the Institute of General and Inorganic Chemistry), on the ozonization of metallic calcium–liquid NH_3 solutions, indicated that the solutions acquire a bright orange color which fades away after the ammonia is evaporated. The deposit consists mainly of the metal ammoniate. The discovery of ammonium ozonide [34] makes it possible to attribute the color of the ozonated alkaline earth metal–liquid ammonia solutions to the formation of this unstable compound. The function of the metal is limited to the liberation of ammonium ions. In the case of calcium, the reactions are assumed to proceed as follows:

$$10NH_3 + 6O_3 + 3Ca \rightarrow Ca \cdot 6NH_3 + Ca(OH)_2$$

$$+ Ca(NO_3)_2 + 2NH_4O_3 + 2O_2 + H_2$$

REFERENCES

1. I. A. Kazarnovskii and G. P. Nikol'skii. Dokl. Akad. Nauk SSSR 64:69 (1949).
2. G. P. Nikol'skii, Z. A. Bagdasar'yan, and I. A. Kazarnovskii. Dokl. Akad. Nauk SSSR 77:67 (1951).
3. E. Würtz. Dictionnaire de Chimie Pure et Applique, Vol. 2 (1868), p. 721.
4. A. Bayer and V. Villiger. Ber. 35:3038 (1902).
5. W. Strecker and H. Thienemann. Ber. 40:4984 (1907).
6. G. S. Zhdanov and Z. V. Zvonkova. Zh. Fiz. Khim. 1:100 (1951).
7. H. McLachland, M. Symons, and M. Townsend. J. Chem. Soc. (1959), p. 953.
8. A. K. Piskunov, A. A. Manenkov, and Z. A. Bagdasar'yan. Zh. Eksperim. i Teor. Fiz. 37:302 (1959).
9. J. E. Bennett, D. J. E. Ingram, and D. Schonland. Proc. Phys. Soc. A69:556 (1956).
10. J. A. Marshall and D. O. Van Osterburg. Phys. Rev. 117:712 (1960).
11. I. A. Kazarnoskii. Abstracts of the Eighth Mendeleev Meeting on Pure and Applied Chemistry, Vol. 1, Moscow, Izd. Akad. Nauk SSSR (1959), p. 18.
12. G. M. Schwab and G. Hartman, Z. Phys. Chem. (N. F.) 6:60 (1956).
13. F. Fredenhagen. Z. Anorg. Allgem. Chem. 128:1 (1927).
14. M. Skessarewsky and N. Tchitchinadze. J. Chem. Phys. 14:164 (1916).
15. T. P. Whaley and J. Kleinberg. J. Am. Chem. Soc. 73:79 (1951).
16. I. J. Solomon and A. J. Kacmarek. J. Phys. Chem. 64:168 (1960).
17. I. J. Solomon and A. J. Kacmarek. J. Am. Chem. Soc. 82:5640 (1960).

18. S. Z. Makarov and E. I. Sokovnin. Dokl. Akad. Nauk SSSR 135:606 (1960).
19. V. G. Karpenko and A. S. Poteryaiko. Chemistry of Peroxide Compounds, Moscow, Izd. Akad. Nauk SSSR (1963), p. 193.
20. A. J. Kacmarek, J. M. McDonough, and I. J. Solomon. Inorg. Chem. 1:659 (1962).
21. T. R. Griffiths, K. A. K. Lotte, and M. C. R. Symons. Anal. Chem. 31:1338 (1959).
22. S. A. Tokareva. Chemistry of Peroxide Compounds, Moscow, Izd. Akad. Nauk SSSR (1963), p. 188; Izv. Akad. Nauk SSSR, Ser. Khim. (1964), p. 739.
23a. V. G. Kuznetsov, V. M. Bakulina, S. A. Tokareva, and A. N. Zimina. Zh. Strukt. 7:967 (1962).
23a. V. G. Kuznetsov, V. M. Bakulina, S, A. Tokareva, and A. N. Zimina. Zh. Strukt. Khim. 5:142 (1964).
23b. V. I. Sokol, S. A. Tokareva, and E. I. Sokovnin. Izv. Akad. Nauk SSSR, Ser. Khim. (1963), p. 2220.
24. S. A. Tokareva. Candidate Dissertation, Moscow, IONkh Akad. Nauk SSSR (1964); Izv. Akad. Nauk SSSR, Ser. Khim. (1964), p. 740.
25. N. A. Milas and S. M. Djokic. Chem. Ind. (1962), p. 405.
26. N. A. Milas and S. M. Djokic. J. Am. Chem. Soc. 84:3088 (1962).
27. I. I. Vol'nov, E. I. Sokovnin, and V. V. Matveev. Izv. Akad. Nauk SSSR, Otd. Khim. Nauk (1962), p. 1127.
28. F. Fichter and W. Gladergroen. Helv. Chem. Acta 10:549 (1927).
29. E. Riesenfeld and B. Reinhold. Ber. 42:2180 (1904).
30. F. Mahieux. Bull. Soc. Chim. France (1962), p. 2.
31. I. A. Kazarnovskii and S. I. Raikhshtein. Dokl. Akad. Nauk SSSR 108:641 (1956).
32. E. I. Sokovnin. Candidate Dissertation, Moscow, IONkh Akad. Nauk SSSR (1962); Izv. Akad. Nauk SSSR, OKhN (1963), p. 181.
33. I. A. Kazarnovskii and S. I. Raikhshtein. Dokl. Akad. Nauk SSSR 123:475 (1958).
34. I. J. Solomon, K. Hattori, A. J. Kacmarek, G. M. Platz, and M. J. Klein. J. Am. Chem. Soc. 84:34 (1962).
35. R. Hughes. J. Chem. Phys. 24:131 (1956).
36. P. Smith. J. Phys. Chem. 60:1471 (1956).
37. L. V. Azaroff and J. Corvin. Proc. Natl. Acad. Sci. USA 49:1 (1963).
38. G. P. Nikol'skii and I. A. Kazarnovskii. Dokl. Akad. Nauk SSSR 72:713 (1950).
39. S. Z. Makarov and E. I. Sokovnin. Dokl. Akad. Nauk SSSR 137:612 (1961).
40. I. A. Kazarnovskii. The First All-Union Conference on Ozone, Thesis Report, Izd. MGU (1960), p. 24.
41. I. A. Kazarnovskii and N. P. Lipikhin. Zh. Fiz. Khim. 30:1429 (1956).
42. I. A. Kazarnovskii and N. P. Lipikhin. Theory of Chemical Structure, Kinetics and Reactivity, Reports, Riga (1961), p. 118.
43. I. A. Kazarnovskii, N. P. Lipikhin, and S. V. Kozlov. Izv. Akad. Nauk SSSR, Otd. Khim. Nauk (1963), p. 956.
43a. A. G. Streng. Chem. Rev. 63:620 (1963).
44. I. I. Vol'nov and V. V. Matveev. Izv. Akad Nauk SSSR, Otd. Khim. Nauk (1963), p. 1136.
45. S. N. Foner. In: Free Radicals in Inorganic Chemistry, Advances in Chemistry, Ser. 36, A. C. S., Washington (1962).
46. Chemistry Handbook, Vol. 3, Moscow–Leningrad, Goskhimizdat (1952), pp. 64, 67.
47. E. Manchotte and B. Kampschulte. Ber. 40:4987 (1907).
48. M. Blumenthal. Roczniki Chem. 13:6 (1933); Bull. Intern. Acad. Pol. Sci., A (1935), p. 543.
49. A. Guntz and B. Mentrel. Bull. Soc. Chim. France 29(3):585 (1903).
50. J. Thompson and J. Kleinberg. J. Am. Chem. Soc. 73:1244 (1951).

Conclusion

The formation of peroxide-type compounds with an ionic bond between the element atoms and the molecular oxygen ions, O_2^{2-}, O_2^-, O_3^-, is typical for that small number of elements in the periodic table which have distinctly pronounced basic properties. Elements with distinctly pronounced acid-forming properties produce other types of peroxide compounds, mainly complex.

The abundant number of publications that have appeared in a relatively short time indicates that the most valuable property of peroxides, superoxides, and ozonides is that by which they can be made to readily give off chemically bound oxygen in the active form. This important property draws the ever increasing attention of specialists and technologists. Most of the publications have been dedicated to the problems of synthesis, structure, and thermodynamics. It is felt that as the chemistry of these compounds continues to be studied, new and wider applications will be uncovered for peroxides, superoxides, and ozonides in the various processes of chemical technology.

Index